Animals & Men #46

Typeset by Jonathan Downes,
Cover and Layout by MisterOrange for CFZ Communications
Using Microsoft Word 2000, Microsoft , Publisher 2000, Adobe Photoshop CS.

Photographs © 2009 CFZ except where noted

First published in Great Britain by CFZ Press. Animals & Men is the quarterly journal of the
Centre for Fortean Zoology; a non profit making organisation administered by:

The Centre for Fortean Zoology
Myrtle Cottage
Woolsery
Bideford
North Devon
EX39 5QR

© CFZ MMIX

ISBN: 978-1-905723-41-6

CONTENTS

SUBSCRIPTIONS

For a 4-issue (one year) subscription:

£16 UK £18 EC
£30 US, Canada, Oz, NZ (airmail)
£35 Rest of World.

METHODS OF PAYMENT

Subscription rates INCLUDE postage.

Payment can be made in UK cash, Euro-cheque, or a cheque drawn on a UK bank. Cheques payable to `CFZ Trust`
Britain is one of the few countries in the world where US dollars do not circulate. If making payment in US$ then please add $14 to cover the currency exchange fee.

Payment by credit/debit card through our Paypal account (jon@eclipse.co.uk). Please make all cheques payable to Jonathan Downes

The CFZ Trust is registered as a non profit making organisation with HM Stamp Office. The trustees are J.Downes; R. Freeman & G. Inglis. Charitable status is pending

ABOUT THE CFZ

The Centre for Fortean Zoology [CFZ] is a non profit making organisation which was founded in 1992. It is based in Woolsery, a small village in North Devon, from which we operate, and where we not only have our headquarters, library and museum, but hold our annual convention, *The Weird Weekend*.

After 13 years of being based in Exeter, the CFZ translocated to Woolsery in the summer of 2005, when CFZ Director Jon Downes's father became terminally ill, and the entire Permanent Directorate moved to Woolsery to look after him.

In the seventeen years since the CFZ was formed in the late spring of 1992, it has become the largest and fastest growing mystery animal research group in the world. The CFZ is, however, unique because not only does it carry out research all over the world, but it is active in education and animal welfare work as well.

In the last few years we have become more and more political, because in a world where idiocy is celebrated, and anything of substance is perceived as having less and less value, the CFZ have become more and more needed.

Because it is not just about cryptozoology; it is about utilising the philosophy of Charles Fort, and the methodology extrapolated from it in order to make sense of a natural world which gets ever more complicated, and ever more confusing. That is what we do, and I think that on the whole we do it rather well.

EDITORIAL

Dear friends,

As you have almost certainly noticed, after sticking to basically the same format since we started in 1994, and having been completely the same format since 2003, we have changed completely. This new format of *Animals & Men* has been in the pipeline for a long time, but for various reasons we have always decided to maintain the *status quo*.

However, after we failed dismally to retrieve any semblance of a relationship with the people who have been leasing us printers for the past decade, after their service got worse and worse, and they became the sort of capitalists that you sincerely hope are going to go belly-up during the current recession. We never wanted to run a print shop anyway. It was not one of my better ideas, and we all let out a resounding huzzah as the hairy handed sons of toil removed the bloody thing out of the CFZ front gates, and we steadfastly refused to take delivery of the excessively expensive replacement that they tried to land us with.

There were a few weeks of unpleasantness during which they tried to insist that we pay the remainder of the lease (in complete contravention of the agreement that we had made at the beginning, they have finally just cancelled it all, and we are free of them.

So, we have decided that rather than get another photocopier and continue with the increasingly tedious business of making, collating, and messing about with the bloody things at home, we have bitten the bullet, and decided to turn the magazine into a proper journal and have it printed professionally by Lightning Source who do all our other books.

The changes in the postal system a few years ago also completely banjaxed our pricing system, and we have been waiting until the next format change to rejig our prices and subscriptions, so this seems to be as good a time as any.

For the past six years, *Animals & Men* has been 60pp in length. As of this issue we go up to 96pp (not counting the cover). OK, the last fifteen pages are the current CFZ Book List, and a four page blurb about the

the great days of zoology are not done

CFZ for all the thousands of you readers that I hope will start buying the magazine now it is available on Amazon. There is also a full colour pdf version available much cheaper as a download from the increasingly popular CFZ bloggo. So, finally, we have left our roots in the DIY Fanzine culture of the 1980s, and joined the 21st Century.

About bloody time.

But things are changing all across the CFZ, and mostly for the better. One piece of bad news, however, is that we are no longer involved with Tropiquaria or its proprietors, and so at present are not involved with a zoo. We have no intention of going into any further details, unless it becomes absolutely necessary. Sufficient to say that we have nothing to reproach ourselves for. It is sad that after nearly 15 years Chris Moiser is no longer involved with the CFZ in any capacity, but sometimes that is the way that things work out. I have always said that the CFZ is a big family, and I have always tried to run it as such, but even in the best families one has the occasional divorce.

There have been other changes on the permanent directorate of the CFZ, but nothing of any great importance.

The really important changes have been to our online presence. About six months ago I decided that the CFZ website that Mark and I had set up in 2005 was really getting past its sell-by date. I started work on the new website soon after the Weird Weekend, and really got down to it the day after the last issue of this august journal was dispatched last December. I worked all over Christmas and the new website was launched in the second week of the New Year. However, just at the same time we - pretty well by accident - upgraded what had been my personal blog for the last five years into what has become a daily CFZ magazine with between

fifty and sixty postings a week.

I have to admit that I was a little worried about managing to keep up the levels of original content that we have demanded of ourselves, and also managing to keep *Animals & Men* to the level of content that we have all come to expect, but just in the early stages of putting this edition together I realised that I had nothing to worry about.

At the risk of blowing my own trumpet, I am particularly pleased with this issue, and I think that the editorial team have every reason to feel pleased with themselves.

This is a big year for us. It will see us publish our fiftieth book, it will see the tenth Weird Weekend, and it will see me reach my half century. I had hoped that we would also get to issue 50 of this magazine, but as every cryptozoologist knows, things don't work out as simply as people would like them to. That is what Forteana is all about.

The Centre for Fortean Zoology has changed enormously from the organisation that my ex-wife Alison and I started seventeen years ago, and it is beginning to become the organisation that I always wanted it to be: a truly global resource for researchers and zoological freethinkers across the world.

For the `newbies` to use the unlovely phrase in current parlance, I hope that you like what you see. And for the old guard, some of whom have been subscribers to this magazine since the very first issue back in April 1994, in the immortal words of David St Hubbins: "We hope you like our new direction".

Slainte

Jonathan Downes, Director CFZ

THE FACULTY OF THE CENTRE FOR FORTEAN ZOOLOGY

"In her abnormalities, nature reveals her secrets." (Goethe)

PERMANENT DIRECTORATE

Hon. Life President:
Colonel John Blashford-Snell

****

Director:
Jonathan Downes
Deputy Director:
Graham Inglis
Zoological Director:
Richard Freeman
Administrative Director:
Corinna Downes

****

Ecologist:
Oll Lewis
Technical:
David Braund-Phillips
Aquarist and Invert Specialist:
Max Blake

Local Services Manager:
Matt Osborne
General Services Manager
Tim Matthews
Trainee:
Ross Braund-Phillips
Big Cat Study Group
Neil Arnold

****

North American Office
Nicholas Redfern
Australian Office
Mike Williams
Ruby Lang

****

Asst. Local Services Manager and Assistant to the Director
Emma Biddle
Newsblog Editor
Gavin Lloyd Wilson

Staffordshire:
Lisa Dowley
Surrey:
Nick Smith
Tyneside:
Mike Hallowell
West Midlands:
Dr Karl Shuker
Wiltshire:
Matthew Williams
Yorkshire:
Steve Jones
Yorkshire (South):
Mark Martin

Wales

Elliot Saunders
Oll Lewis
Gwilym Ganes

Northern Ireland

Gary Cunningham
Ronan Coghlan

USA

California:
Greg Bishop
California:
Dianne Hamann
Indiana:
Elizabeth Clem
Illinois:
Jessica Dardeen
Illinois:
Derek Grebner
Missouri:
Kenn Thomas
Ohio:

Chris Kraska
New York State:
Peter Robbins
New York:
Brian Gaugler
New Jersey:
Brian Gaugler
North Carolina:
Shane Lea
North Carolina:
Micah Hanks
Texas:
Chester Moore
Texas:
Naomi West
Texas:
Ken Gerhard
Wisconsin:
Felinda Bullock
Michigan:
Raven Meindel
Ohio:
Brian Parsons
NE Oklahoma:
Melissa Miller
Oregon:
Regan Lee

International

Australia: Tim the Yowie Man
Australia: Mike Williams
Denmark: Lars Thomas
France: Francois de Sarre
Germany: Wolfgang Schmidt
Eire: Tony 'Doc' Shiels
New Zealand: Tony Lucas
Switzerland: Georges Massey
United Arab Emirates:
Heather Mikhail

weird weekend 2009

August 14-16

The world famous event: Three days of monsters, myths, and mysteries. Lectures; workshops; films; music; theatre; exhibitions; stalls. For all ages. Now in its tenth year.

Woolfardisworthy Community Centre, North Devon

www.weirdweekend.org

01237 431413

NEWSFILE

Compiled and Edited by Jon Downes and Richard Freeman

NEW & REDISCOVERED

The largest of the Galapagos Islands is Isabela, known to English explorers as Albemarle. Like so many of the islands it is home to a large lizard, which reaches from 3-5ft in length. Charles Darwin, the zoologist with whom the islands will be forever most closely identified once described them as *"ugly animals, of a yellowish orange beneath, and of a brownish-red colour above: from their low facial angle they have a singularly stupid appearance."*

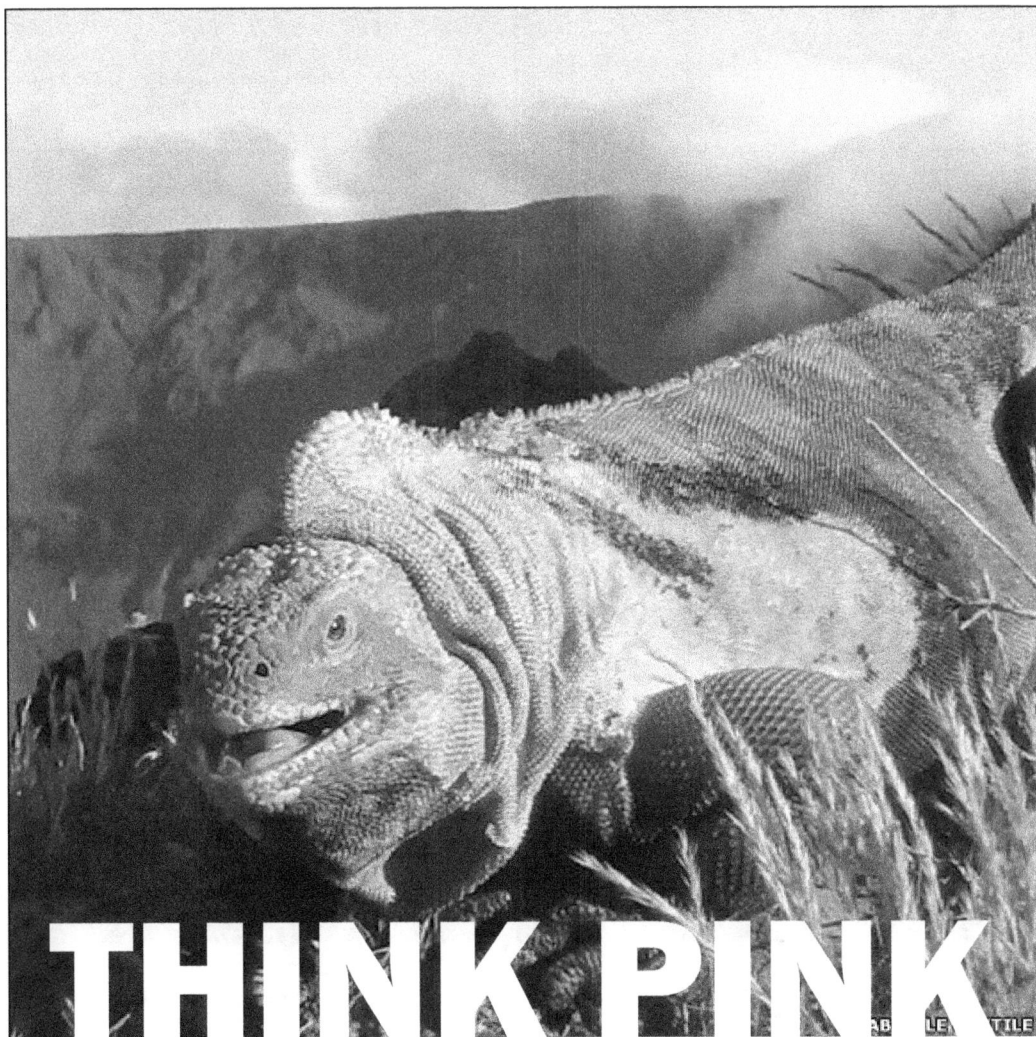

THINK PINK

Until recently it was thought that the genus *Conolophus* had two species, *Conolophus subcristatus*, the Galapagos land iguana, and a lesser known species *Conolophus pallidus* only found on the island of Santa Fe. However, now a third species has been described, and it is the most outstanding of them all. Why? Because it is bright pink in colour!

The new species, *C. rosada* or the pink iguana is only found on Volcán Wolf, the highest mountain on Isabela, or indeed on any of the Galapagos Islands.

It was first discovered in 1986, but it has taken twenty-three years for the lizards to be identified as a separate species. It is ironic that although Isabela is one of the youngest of the Galapagos chain, and as such has a relatively sparse fauna, this split between the newly discovered species and the more well known lizards that Darwin sneered at so resoundingly, is that the evidence suggests that *C. rosada* separated from the main species as long as 5.7 million years ago. *"At 5.7 million years ago, all of the western islands of the archipelago did not exist,"* said Gabriele Gentile from the University of Rome Tor Vergata, who led the new analysis.

"That's a conundrum, because it's now only inhabiting one part of Isabela that formed less than half a million years ago," he told BBC News. And the real conundrum is that even the oldest parts of the archipelago may be less than five million years old.

But the new pink iguana is not the only new iguana to have been discovered in 2009.

A new iguana has been discovered in the central regions of Fiji. The colourful new species, named *Brachylophus bulabula*, joins only two other living Pacific iguana species, one of which is critically endangered. The scientific name bulabula is a doubling of bula, the Fijian word for 'hello,' offering an even more enthusiastic greeting.

Pacific iguanas have almost disappeared as the result of human presence. Two species were eaten to extinction after people arrived nearly 3,000 years ago.

The three living *Brachylophus* iguana species face threats from loss and alteration of their habitat, as well as from feral cats, mongooses and goats that eat iguanas or their food source.

"In the reptile world the Fijian iguanas are iconic," says lead author Associate Professor Scott Keogh, of the Australian National University's School of Botany and Zoology. *"To discover a new species of them is very exciting."*

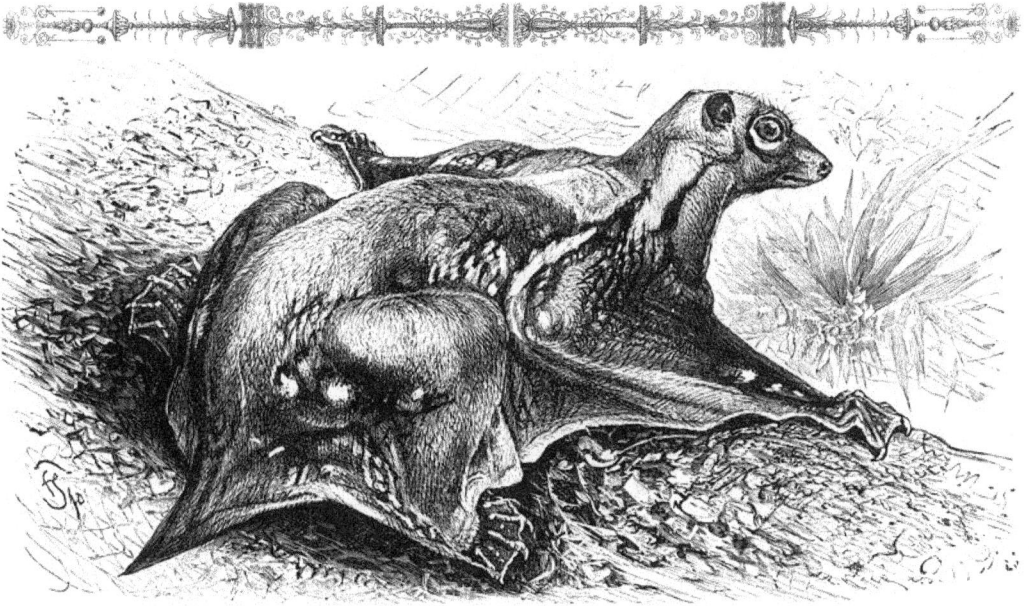

Colugos are peculiar animals from southeast Asia. As a small child I was given a copy of *Mammals of the World* by Hans Hvass, and I became entranced by the more obscure groups of mammals, like pangolins, aardvarks, monotremes and particularly colugos. Colugos or flying lemurs were once considered to be reasonably closely related to bats, but in recent years it has come to be accepted that they are actually something that I had considered to be likely for many years: the closest living relatives of primates, having diverged from that group about 86 millions years ago during the Late Cretaceous.

Until recently, scientists recognized just two colugo species, the Sunda Colugo (*Galeopterus variegatus*) and the Philippine Colugo (*Cynocephalus volans*). But researchers analyzing genetic material from Sunda Colugos living on the Malay Peninsula, Borneo and Java found genetic differences great enough to suggest that the colugos living on each island had evolved into distinct species.

The new distinct species of colugo also look slightly different. For instance, the colugos on Borneo are smaller that their Javan and mainland counterparts and the Borneo colugos also have a wider variation than their relatives in fur colour, including some with spots and others with really dark colouring.

"We were guessing that we might find that there were different species of Sunda colugo-although we were not sure," said Jan Janecka of Texas A&M University. *"But what really surprised us was how old the speciation events were. Some went back four to five million years,"* making the colugo species as old as other modern species groups (or genera) such as the macaques and the leopard cats.

The team's initial hunch that the Sunda colugos might be distinct species came largely from obvious differences in characteristics like body size and colour. In the new study, they compared the DNA of colugos living on the mainland, Java, and Borneo, uncovering enough divergence between the sequences to warrant their designation as three species.

Janecka said they were particularly surprised to find that each geographic region they studied harbours its own unique species of colugo.

And the species tally for colugos will most probably rise some more. *"It appears that within smaller geographic areas, for example Java, there are divergent colugo lineages that could prove to be separate species,"* he added.

COLUGO CONUNDRA

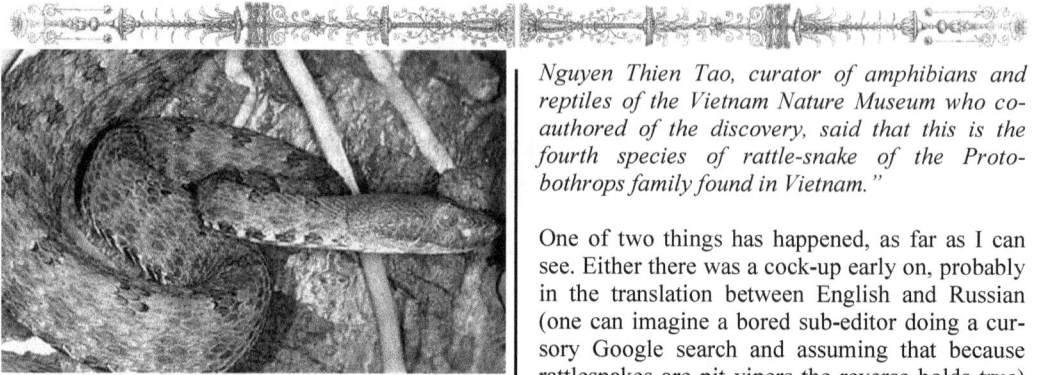

There is a new species of Vietnamese pit viper (*Protobothrops trungkhanhensis*). That's jolly nice, but the case actually presents some interesting Fortean characteristics, because for some reason known only to the Gods of the internet, every single news item that we have found about the snake describes it erroneously as a brand new species of "rattlesnake".

Well rattlesnakes are indeed a subgrouping of pit vipers, but whereas all rattlesnakes are pit vipers, not all pit vipers are rattlesnakes, and not only is the new species not a rattlesnake, all rattlesnake species are found exclusively in the New World.

The thing that is both interesting and worrying is the way that the completely erroneous information has been disseminated willy nilly across the internet.

Even www.wildlifeextra.com who are a generally excellent source for information write:

"*The newly discovered rattle-snake is named* Protobothrops trungkhanhensis *Orlov, Ryabov & Nguyen. The snake is relatively small when compared to other* Protobothrops *species, being only 733mm in length, with a small triangle-shaped head.*

Nguyen Thien Tao, curator of amphibians and reptiles of the Vietnam Nature Museum who co-authored of the discovery, said that this is the fourth species of rattle-snake of the Protobothrops family found in Vietnam."

One of two things has happened, as far as I can see. Either there was a cock-up early on, probably in the translation between English and Russian (one can imagine a bored sub-editor doing a cursory Google search and assuming that because rattlesnakes are pit vipers the reverse holds true) or - more disturbingly - there is an entire culture of journalists just cutting and pasting other people's information.

Now, I am very aware of the Biblical quotation about `casting the first stone` at this point. Some years ago, during one of my more spectacular bouts of uncontrolled bipolar activity, I had a job working for a natural history partwork.

I blotted my copybook seriously by resorting to the aforementioned cut and paste techniques, and was sacked. In my defence I was as mad as a bagful of cheese at the time and I have no real recollection of the events in question, but I will freely put my hands up and say *mea culpa.*

But that was long before I had any pretensions to being a reputable scientific journalist, and the fact that the rattlesnake identification has been used on so many occasions to describe this totally un-rattlesnake (it doesn't even have a rattle for goodness sake) is a little disturbing.

The other explanation is that there has been a major re-evaluation of the taxonomy of pit vipers, and both Richard's and my information is out of date, and we are just making idiots of ourselves by making an issue of this case. It is certainly a possibility.

IT'S THE PITS

For a cryptid that may well be an undescribed species of pit viper go to page 25

WORCESTER'S BUTTONQUAIL
(*Turnix worcesteri*)

A rare Worcester's buttonquail, locally known as Philippines quail, is photographed while being held by a bird hunter in Caraballo mountain, on Luzon island, in January, after it was snared in a trap. The bird was previously only known through drawings based on dead museum specimens collected decades ago, and has not been reported alive for a century, the Wild Bird Club of the Philippines (WBCP) has reported.

Unfortunately, the bird was eaten shortly after being photographed as no one at the time realised the bird's importance. The WBCP hailed the discovery of the Worcester's buttonquail. *"We are ecstatic that this rarely seen species was photographed by accident. What if this was the last of its species?"* WBCP president Mike Lu said.

GILES'S ANTPITTA
(*Grallaria milleri gilesi*)

Extinction is a strange concept; and for a word which is used so often these days, many people don't actually sit down and work out what it actually means. There are often peculiar stories surrounding the extinction of a particular species, and the stories are additionally poignant, when they also involve the discovery. Take this little bird from Colombia, for example. It is about the size of a thrush, it is flightless, it is small and brown, and looks almost identical to a well-known bird pictured on its right.

It is a brand-new subspecies, and it was discovered amongst the million bird specimens at the British Museum of Natural Nistory in London. The peculiar part of the story is that it had languished unrecognised in the collections for 120 years, and even though it was finally given subspecific status in March, it is almost certainly extinct, and probably has been for years.

Bird artist Norman Arlott uncovered the specimen in 2001 and drew it to the attention of Museum's Head of Bird Group, Dr Robert Prys-Jones, and Museum scientific associate Dr Paul Salaman. *'We immediately recognised the specimen was previously undescribed,'* says Dr Salaman, *'yet it seems inconceivable that this distinctive specimen could have had been overlooked for 120 years.'* The bird was originally collected in September 1878 by the British ornithologist Thomas Knight Salmon.

MAGDALENA VALLEY RINGLET
(*Splendeuptychia ackeryi*)

Another new species, with a "does it/does it not" status reminiscent of Schrödinger's eponymous felid is another species from Colombia - butterfly which again was discovered in the hallowed archives of the British Museum (Natural History) in South Kensington.

It was discovered by Blanca Huertas, discovered by a curator from The Natural History Museum. Huertas collected *S. ackeryi* during an expedition to Columbia, but it wasn't until the unidentified butterfly was compared with a museum specimen that entomologists realised it

NEW AND REDISCOVERED IN BRIEF

was the same as a 90 year old unidentified museum specimen. One of the distinguishing features of *S. ackeryi* which helped with the identification were its extremely hairy mouthparts.

WORLD'S OLDEST ANT
(Martialis heureka)

Most scientific discoveries these days emerge through carefully planned and controlled research programs. Every now and again, though, something unexpected just pops up in a distant tropical jungle. *Martialis heureka* is a fantastic discovery of that old-fashioned kind. This little ant simply walked up to myrmecologist Christian Rabeling in the Brazilian Amazon. It is not only a new species, but an entirely different sort of ant than anything known before.

Two specimens were first discovered by Manfred Verhaagh of the Staatliches Museum für Naturkunde in Karlsruhe, Germany in 2000, but they became desiccated and were broken on their way to being tested, and so could not be verified as being a new species. It wasn't until another specimen was found in 2003 by Christian Rabeling, a graduate student of the University of Texas at Austin, that the discovery finally came to light. Known only from these three specimens, the species is placed in a monotypic genus and a new subfamily (**Martialinae**) that is a sister lineage to all other living ants and which is the first extant ant subfamily to be discovered since 1923.

The ants themselves are eyeless, pale in colour, subterranean, and predatory. This doesn't mean that the ancestor to all ants was blind and lived underground, but that these features evolved early in ant history and have persisted in the environmentally stable soils of the tropical rainforest.

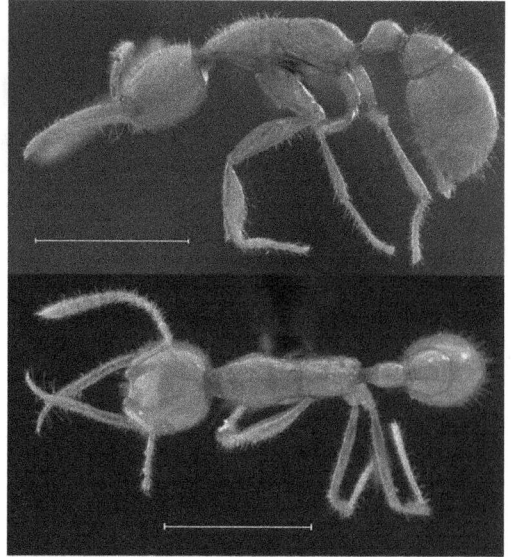

PSYCHEDELIC WALKING FROGFISH
(Histiophryne psychedelica)

As readers of this magazine, and my inky-fingered scribblings in other locations will know, I was once a rare record dealer, and specialised in collecting, ummmmmm how do I put this, "unofficial" recordings by a certain four piece beat combo from Liverpool called *The Beatles.* There is a song which for years eluded me called *Oriental Nightfish.* It was written by Linda McCartney, recorded during the sessions for the 1973 *Band on the Run* album by Paul McCartney's post-fabs band *Wings,* and to be quite honest it is not terribly good. In fact it is terrible, and living proof that there is no limit to how far you can get on the barest modicum of talent, when you are married to a superstar. How-

NEW AND REDISCOVERED IN BRIEF

harbour of Ambon Island, Indonesia. An adult fish was observed in January 2008 by Toby Fadirsyair, a guide, and Buck and Fitrie Randolph, two of the co-owners of Maluku Divers, which is based in Ambon. They and co-owners Andy and Kerry Shorten eventually found Pietsch to help them identify the fish. Since the first sighting divers have observed a number of adults and juveniles, now that they know what to look for.

AMPHIBIAN EXPLOSION

So far this has been a remarkable year for new amphibian species. Early in the year ten new frog species were announced from Sri Lanka, and another ten amphibia from Colombia, but the news in early May took the figurative biscuit. *Science Daily* wrote:

Between 129 and 221 new species of frogs have been identified in Madagascar, practically doubling the currently known amphibian fauna. The finding suggests that the number of amphibian species in Madagascar, one of the world's bio-

diversity hotspots, has been significantly under-estimated. According to the researchers, if these results are extrapolated at a global scale, the number of amphibian species worldwide could double.

This just goes to show that we do only know a fraction about what the natural world has to show us.

ever, for the last few months the bloody thing has been going round and round my head, and it is all because of a new species of walking frogfish discovered in Indonesia.

The new species gets its name from its swirling colours of peach and blue, but their singular mode of propulsion also reminds one irresistibly of a creature seriously under the influence of mind-altering chemicals. *H. psychedelica*, don't so much swim as hop. Each time they strike the sea-floor they use their fins to push off and they expel water from tiny gill openings on their sides to jettison themselves forward. With tails curled tightly to one side – which surely limits their ability to steer – they look like inflated rubber balls bouncing hither and thither.

While other frogfish and similar species are known to jettison themselves up off the bottom before they begin swimming, none have been observed hopping quite like this. It's just one of the behaviours of *H. psychedelica* never observed in any other fish, says Pietsch, UW professor of aquatic and fishery sciences and curator of fishes at the UW Burke Museum of Natural History and Culture. He's the lead author of a paper about the new species that's now online at *Copeia*, the journal of the American Society of Ichthyologists and Herpetologists. His work is funded by the National Science Foundation.

It was little more than a year ago that the fish with rare, forward-facing eyes like humans and a secretive nature was the subject of worldwide news coverage after having been observed in the busy

NEW AND REDISCOVERED IN BRIEF

ARBOREAL GECKO
Ptychozoon nicobarensis

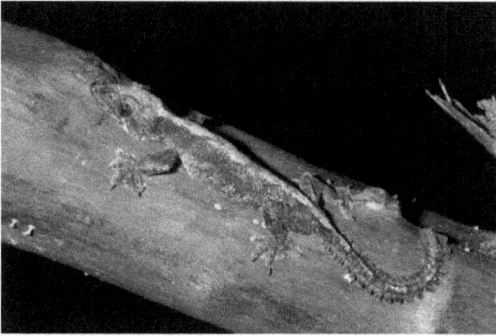

A new species of arboreal gecko has been discovered in the Nicobar Archipelago of the Bay of Bombay, India. The species, *Ptychozoon nicobarensis*, was previously thought to be a variant of *P. kuhli*, and grows to around 10cm (4ins) in length. The geckos are only found in the central portion of the archipelago and scientists speculate that this is due to competition and predation from large gekkonid species in the area, *Gekko verreauxi* and *G. smithii* to the north and south respectively.

CUBAN CROAKING GECKO
Aristelliger reyesi

A croaking gecko species from the genus *Aristel-*

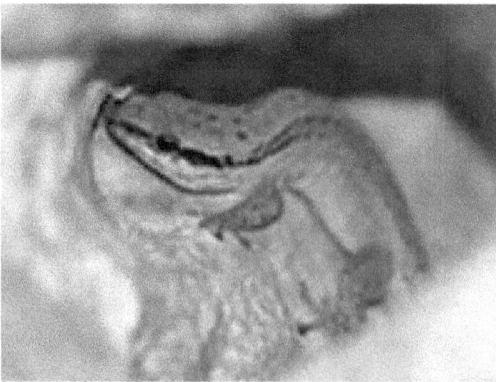

liger, has been indentified on the island of Cuba for the first time. The newly discovered species, *Aristelliger reyesi,* shares common morphological features with the subgenus *Aristelligella* including fragile skin, friction pads on it's feet and a supraciliar scale modified to form a small spine, but has a different colour pattern. Other species of *Aristelliger* are found throughout the Caribbean and Central America.

TWO MOUNTAIN HORNED DRAGONS

Acanthosaura bintangensis

A. titiwangsaensis

Two new species of mountain horned lizards from the genus *Acanthosaura*, were discovered in Malasia. Both species closely resemble each other the only differences being *A. bintangensis* having more subdigital lamellae on it's fourth fingers than *A. titiwangsaensis*, larger suborbital

NEW AND REDISCOVERED IN BRIEF

scales, more spots on it's back, and by having a black gular region with a yellow medial stripe.

NEW BEETLE SPECIES
Leptogenopapus mirabilis

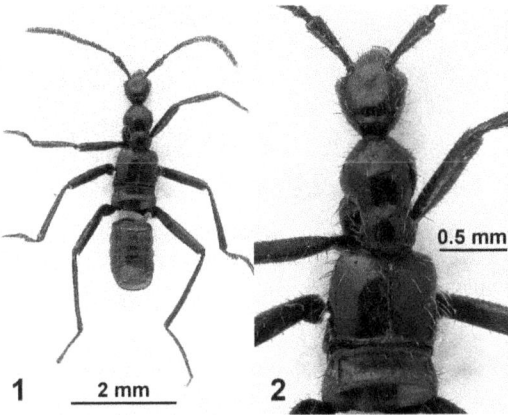

1   **2 mm**   **2**   **0.5 mm**

A new genus and species of beetle, *Leptogenopapus mirabilis*, has been identified in Papua New Guinea. The beetle was found with a foraging colony of *Leptogenys breviceps* ants and the species is thought to have a close relationship with them like other staphylinid beetles. *L. mirabilis's* holotype is male and has a body length of 3.8mm and a width of 1.25mm and is light brown in colour.

NEW NORTH AMERICAN WINDOW FLY
Pseudatrichia bezarki

A previously un-catalogued species of 'window fly', *Pseudatrichia bezarki*, has been identified in Northern Arizona. Adults of both sexes measure around 5mm in length and are gloss black in colour. *P. bezarki* is thought to be closely related to *P. punctulata* and *P. rufitruncula*.

Dampierella schwartzi

Goodeniaphila cassisi   *Goodeniaphila schuhi*

TWO NEW GENERA AND THREE NEW SPECIES OF HALTICINI (PLANT BUGS)
Dampierella schwartzi, Goodeniaphila cassis, G. schuhi

Three new species of Halticini, *Dampierella schwartzi*, *Goodeniaphila cassis* and *G. schuhi*, have been discovered in Australia.

Both of the new genera were based upon the

NEW AND REDISCOVERED IN BRIEF

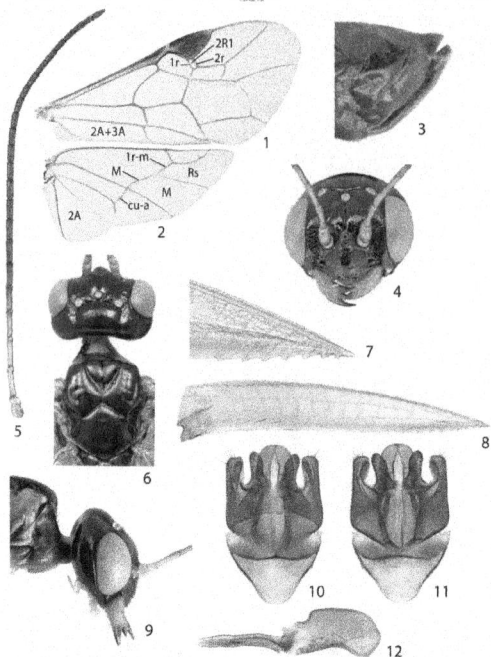

plants that the species are associated with; Plants of the Goodeniaceae family in the case of *Goodeniaphila* and *Dampiera incana* var. *incana* in the case of *Dampierell.*

AUSTRALIAN CEPHIDAE
Australcephus storeyi

A new subfamily, genus, and species of Cephidae or stem sawflies has been identified in Australia, based on the newly discovered species *Australcephus storeyi* in Queensland.

A. storeyi is one of only three species of Cephidae now known to be native to the Southern Hemisphere. Both males and females measure around 5mm in length.

INDIAN OCEAN ECHINODERM
Gebrukothuria profundus

A new species and genus of laetmogonid holothurian has been discovered on the Crozet Plateau in the Southern Indian Ocean. It is the only member of the Laetmogonidae family that lacks wheel-shaped calcareous deposits in its body wall completely.

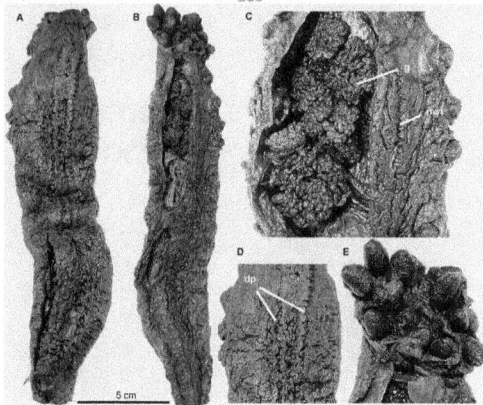

Because the calcareous deposits were previously one of the diagnosing factors of the Laetmogonidae family the family description has had to be altered to take account of this.

NEW DARTER
Etheostoma erythrozonum

A new species of darter, *Etheostoma erythrozonum*, is described from the Meramec River drainage of Missouri, USA.

Etheostoma erythrozonum is the first known fish

FIGURE 1. *Etheostoma erythrozonum* (A) male 67 mm SL, holotype USNM 391646 and (B) female 64 mm SL, paratype USNM 391647. Huzzah Creek at the Reis Biological Station, 6.2 km upstream from the Route 8 bridge, Crawford County, Missouri, 3 April 2003.

NEW AND REDISCOVERED IN BRIEF

species endemic to the Meramec River drainage. It differs morphologically and genetically from populations of its sister species, *Etheostoma tetrazonum*, from the Gasconade River, Osage River, and Moreau River drainages.

NEW ELECTRIC FISH
Brachyhypopomus gauderio

Brachyhypopomus gauderio is described here from the central, southern and coastal regions of the Rio Grande do Sul state, Brazil, from Uruguay, and from Paraguay. It is diagnosed from the congeners on the basis of body coloration, the number of anal-fin rays, the position of anal-fin origin in relation to pectoral-fin, the morphology of the distal portion of caudal filament of mature males, and body proportions.

Diversity of the genus *Brachyhypopomus* Mago-Leccia is clearly underestimated. All recent reviews of the species composition of the genus have resulted in a list of six species in this group of Neotropical knifefish (Mago-Leccia 1994; Albert 2001; Albert & Crampton 2003), while studies published just in the last five years have increased this number by 50 percent.

FIVE NEW DAMSELFISH

Five new species of the damselfish genus have been described from specimens collected from deep (>60 m) coral-reef habitat in the western Pacific by divers using mixed-gas closed-circuit rebreather gear. Two of the five new species (*Chromis. abyssus* and *C. circumaurea*) are each described from specimens taken at a single locality within the Caroline Islands (Palau and Yap,

respectively); one (*C. degruyi*) is described from specimens collected or observed throughout the Caroline Islands, and two (*C. brevirostris* and *C. earina*) are described from specimens collected from several localities throughout the Caroline Islands, Fiji, and Vanuatu. All five species can easily be distinguished from other known *Chromis*, and from each other, on the basis of colour and morphology.

Chromis abyssus

Chromis brevirostris

Chromis circumaurea

NEW AND REDISCOVERED IN BRIEF

HOAXES

This picture was printed in an Australian newspaper, *The North West Star* [Mount Isa, Queensland] on the 9 April 2009, under the headline *Another urban myth?*

I am not going to display the level of pedantry shown by Bernard Wooley in *Yes Minister*, and say that this could not be another "urban" myth as it purports to have taken place in the countryside, because there are far more important implications to this story than mere point scoring.

The story reads:

SOME Doomadgee residents have dubbed it the "Water Blackfella" but we at The North West Star are not so certain. A couple came into our offices this week claiming to have an image of a mysterious creature sighted in the Gulf of Carpentaria.

According to the legendary tale being told around kitchen tables in the region, a Doomadgee man was trying to take a photo of a large snake at Drum Yard Station, which is about 50 kilometres outside Doomadgee, when this creature emerged from a billabong. The quick thinking man supposedly hid behind a tree and took this "photograph" before the monster slinked past him and out of sight.

The picture is currently being passed from phone to phone throughout Doomadgee as residents debate its authenticity. Is it bunyip, a gorilla, a water blackfella or just simply (and most likely) a hoax? The picture looks pretty suss to us, but we'll let you decide - believe it or not.

Well, the unnamed Domadgee man may well have seen a giant unknown hominid, although we strongly doubt it. But there is no doubt that the photograph is not of a "water blackfella" or anything like it, and that it certainly wasn't taken in April 2009.

On the ever-evolving CFZ bloggo we have been tracing back this picture, and the earliest appearance that we have found for it is back in the early 1990s in *Fate* magazine.

One of the most intriguing things about the whole affair is that anyone ever took this picture seriously in the first place. However, the same thing could well be said about the Mary F photos, the 1963 Tim Dinsdale movie, and various other icons of cryptozoology which - let's admit it - do nothing but illustrate the enduring power of self-delusion in people who so badly "want to believe".

Bloggo stalwart, and relatively new member of the CFZ family, Glen Vaudrey saw the story, and if I may quote what Penny Rimbaud said about the time in early 1977 when he and Steve Ignorant saw *The Clash* for the first time, he saw it as a "challenge to his creativity". So, he set out to see if he could replicate the pictures and construct a "water blackfella" of his own.

He writes:

"Jon,

I sent you an email a couple of weeks ago bragging that I could produce some made at home Blackfella pictures, well here are the pictures.

Sorry no great cryptozoological breakthrough involved just two quid's worth of plasticene, a piece of thick wire and plenty of spare hair left after my last haircut.

Maybe not the most convincing images but near enough for a quick glance".

I think that Glen has proved his point admirably.

But this wasn't the last hoax to be perpetrated during the long, and exceptionally balmy (in both senses of the word) spring of 2009.

Something that I found interesting was the racism inherent in the name of this otherwise unknown Australian non-cryptid. Australia is as hidebound with political correctness as we are in the UK, but it is hard to imagine any regional paper in the UK using such a term, although - much to my surprise - according to Wikipedia:

Blackfella (also spelled blackfellah, black fella, or black fellah) is an informal term used in Australian English to refer to Indigenous Australians. It derives from "black fellow," referring to their dark skin. It is generally considered to be a neutral term, and is used by both white and black Australians.

But the most important thing about this story is, that although I have never heard the term before, it is strikingly similar to a South American term "Negroes of the Water" used to describe malicious small humaniform entities, including some described in Charles Bowen's seminal 1969 book *The Humanoids* which is sadly out of print.

It has been quite a bumper year so far for hoaxes. There has been a spoof sea-serpent in the English Channel, there was the `Water Blackfella` saga described on the previous two pages, and there was a particularly pointless story which came from the Gulf state of Quatar.

On the 23rd April the *Gulf Times* reported:

"A mysterious figure resembling a human being was sighted on the Doha Corniche's parking lot, according to a report published in a local Arabic daily. The report is based on the statement of an Arab expatriate lady who said she had seen the strange figure near the Oryx statue while walking in the area. Quoting the woman, the daily said she took a picture of it in spite of being terribly frightened. "She was very soon surrounded by a large number of people who also attested to the fact of what she had seen . But it suddenly disappeared out of their sight when they tried to go near it," the report added."

It was obviously a hoax, and I made some facile joke about Adobe Photoshop and Salman Rushdie on the bloggo and then forgot about it.

However, a few days later, Mark North (God bless 'im) e-mailed us a photograph of the true culprit:

I may have found your answer, it is indeed a cheap stretchy rubber toy that can be found in many shops including Hawkins Bizarre where I last saw one, and thought it was the worst depiction of a werewolf I had seen since watching 'Werewolves of the Blood Moon'. Check it out http://www.stocking-fillers.co.uk/find/product-is-07917?img=_d

So, as I wrote on the bloggo at the time, the moral of this story is that, in cryptozoological terms as in everything else in life, it is wise to look before you leap.

However, I was not prepared for a backlash. I suppose that I should have been because pretty well everything we do results in a bloody backlash. However, I never for a moment thought that writing disparagingly about a rubber model would get torrents of disapproval. But as my ur-hero Lazarus Long is mooted to have said, one should never underestimate the power of human stupidity.

I have been accused, both overtly and covertly, of being a traitor to cryptozoology by "refusing to investigate incidents that don't fit in with my world view".

Read my lips: It was a rubber toy. Any moron would have seen that it was a rubber toy, and furthermore, far from refusing to investigate the matter, we *did* investigate the matter, and furthermore we *solved* it.

It is about time that cryptozoology as a discipline grew up. To give credence to palpable hoaxes like this is the mark of the deluded wannabe. If cryptozoology is to be taken seriously then it has to develop a sense of proportion, and to exercise a degree of quality control onto the subjects with which it deals.

Here endeth the rant, and I am getting off my soapbox now.

XTRA NEWS FILE

Tsuchinoko: An un-known Pit Viper?

The 'flat snake', is a yokai that may have it's roots in reality. *Tsuchinoko* is reported to be between one - three feet long with a dorso-ventrally compressed body. It has horn like ridges above the eyes and dorsal pits. The head is triangular and the neck well defined. It is thought to be venomous. It is described as being black or rust coloured. It is supposed to have an odour like chestnut tree flowers!

In legend it has a liking for alcohol, can speak and can hold its tail in its mouth, and roll along like a hoop (such stories are common in western snake legends as well). It is also said to progress in a series of hopping motions.

The creature is known as *Bachi hebi* in Northeastern Japan and also *tsuchi-hebi*. It is also said to live in Korea. This yokai has a long history. Drawings resembling *Tsuchinoko* on stoneware dating back to the ancient Jōmon Period (14,000 BC to 300 BC.) have been discovered in Gifu and Nagano. An encyclopedia from the Edo Period contains a description of the tsuchinoko under the name *Yatsui hebi*. Accounts of the Tsuchinoko can also be found in the Kojiki, Japan's oldest surviving book from 680 AD.

But this is a yokai that is still reported in modern times. In June, 1994, 73 year old Kazuaki Noda was cutting grass with his wife when they came across a huge snake with a thick body like a beer bottle, and a head described as being like that of a tortoise.

On May 8th, 2000, 90 year-old farmer Sugie Tanaka was out looking for bamboo shoots in Mikata, Hyogo prefecture, when she happened across two metallic coloured snakes with what she described as "tails like rats."

On May 21st 2000 in Yoshi, Okayama prefecture, a farmer cutting grass saw one of these weird serpents slithering across his field. He said it's face reminded him of the popular Japanese cartoon cat Doraemon. He slashed at it with his weed whipper but it escaped into a stream. He said he had heard of the creatures before and thought they make a kind of chirping sound.

Four days later, 72 year-old Hideko Takashima found the creature's body lying beside a stream. Government officials collected the body and sent it to Kawasaki University of Medical Welfare to be examined. Professor Kuuniyasu Saton examined and said that the creature may indeed have been the beast referred to as *Tsuchinoko* in ancient legends but *"scientifically speaking, it*

was a kind of snake."

If the professor ever established the species or if it was indeed a new species is unknown.

A live *Tsuchinoko* was reportedly captured in Mikata on June 6th, 2000. It was supposedly put on display in a glass box in the Mikata's visitor centre. It may well have been a hoax to drum up publicity for the *Tsuchinoko* hunts held each year in the area.

Also in in June, 2000. 82 year-old Mitsuko Arima saw a *Tsuchinoko* swimming along a river. She described its eyes as being the most striking feature, saying *"I can still see the eyes now. They were big and round and it looked like they were floating on the water."* She added *"I've lived for over 80 years but I've never seen anything like that in my life."*

On 30th of August 2000 the *Mainich Daily News* reported that a bounty had been put on *Tsuchinoko's* head. People throughout Kansai and from as far away as Kanto had throng to the town in central Okayama Prefecture in order to try and catch a *Tsuchinoko* and claim the reward of twenty million yen from the Yoshi Municipal Government. Though the scaly little yokai remained elusive local sales of *Tsuchinoko* wine and *Tsuchinoko* rice cake sky rocketed.

Other supposed bodies of this animal have turned up. In Mikata, (the area said to have the highest concentration of sightings in Japan) a corpse was found by four loggers in the Spring of 2001. The body was actually turned over to the Japan Snake Centre in Gunma prefecture, where an analysis was done on it that confirmed it as a common grass snake. Another body was found by a villager near Mikata around the same time in May that year, and it too was turned in to the same centre for examination. It was determined to be a rat snake.

Back in 1969 A live *Tsuchinoko* was reportedly captured in by an M. To-kutake in Mikata. He supposedly captured it with a forked stick and kept it

for a couple of days before deciding to eat it.

According to Naoki Yamaguchi, who has interviewed over 200 eyewitnesses and is author of the book *Catching the Illusory Tsuchinoko*, the specially organized mass searches are of little use. The searchers do not go far enough into the wilderness. Most sightings are by hikers, fishermen and loggers.

If Tsuchinoko is a real creature then it seems probable to me that it is an aberrant species of pit viper (Crotalinae). The flattening of the body could enable the snake to hide in rock crevices. There is a president for thin in reptiles. The chuckwala (*Sauromalus*) are a genus comprising of five species of Iguanid lizards of the arid areas of the southern USA. They are dorsoventrally flattened and once in a crevice, inflate their bodies to make extraction by a predator near impossible.

The Pancake tortoise *(Malacochersus tornieri)* of eastern Africa has a flattened shell so it can slide into crevices. It is not impossible that Japan, a country with a surprising number of snake species, plays host to an undiscovered pit viper of unique form, and that one of the most ancient yokai is in fact very real.

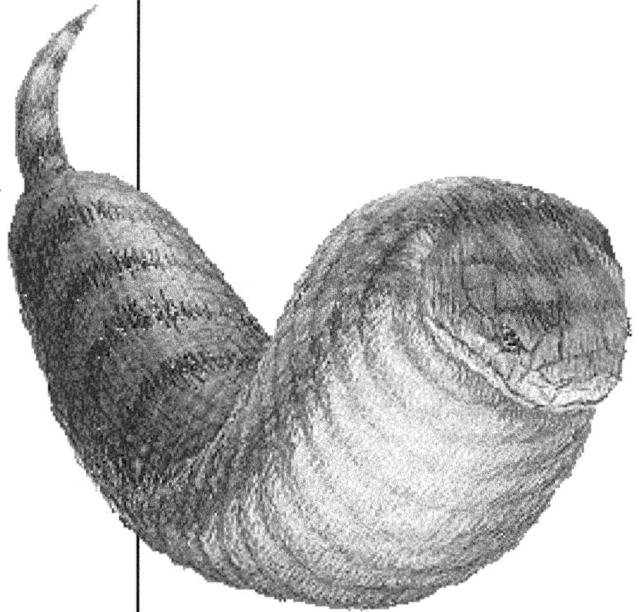

BEAST ON THE BEACH
Riddle as razor-toothed remains are washed up

WHAT COULD MYSTERY MONSTER BE?

The year was only a couple of weeks old when a seemingly pointless, but eventually pivotal event took place. It all started when I received a telephone call from an acquaintance on one of the local newspapers. *"Would I give a comment on the discovery of the 'Beast of the Bay'?"* she asked.

I had no idea what she was talking about, and politely told her so. She told me in a slightly disgruntled manner that a mysterious carcass had washed up on the beach at Croyde Bay, and that because of its enormous teeth, experts had identified the corpse as that of the 'Beast of Exmoor'. Upon being questioned further, it turned out that the 'experts' were two blokes who happened to be walking along the beach at the time. She reluctantly e-mailed me the first pictures of the carcass, with strictures that I was not to reuse, republish or resell them upon pain of dire stuff happening to me.

As soon as the pictures arrived it was obvious what they were. It was a dead pinniped of some description, and I identified it as being "almost certainly that of a grey seal. However, we decided that as part of our quest to obtain a representative collection of skulls of local wildlife for identification purposes, and also because it was a slow day, that Oll, Graham and Matt Osborne should go to Croyde and photograph the specimen in situ and bring back the skull for the collection.

This they did. And it did indeed turn out to be a grey seal, despite the exciting possibility that it could have been a much rarer pinniped, and there the matter would have rested if it had not been for the eccentricities of Her Majesty's Press.

To quote from a statement we released at the time:

Because the next day *The Sun* proclaimed, *"Last night, in a further twist, police reported the beast's skull STOLEN."* Well it hasn't been... we've got it! Because of the possibility that this corpse might conceivably be the body of only the second sealion ever to turn up on British waters, we were con-

... and could THIS be the beast of Exmoor?

Paul Harris reports

Looking beastly: The toothy remains

IT was the teeth that everyone noticed first.

Great fangs jutted from its huge jaw, gleaming in the afternoon sun.

Then there was the carcass. Up to 5ft long, powerful chest, and what could be the remains of a tail.

Had it been washed up on any other shore, it might simply have been dismissed as the unfortunate remains of a large dog.

But this was North Devon. And folk in these parts have learned that sightings of mystery animals are likely to mean only one thing - the Beast of Exmoor is back.

The puma-like creature has allegedly roamed the countryside here since some fleeting glimpses in the 1970s. In 1983, it came to national attention after 100 sheep were mauled and killed. Blurred photographs and a succession of intriguing sightings followed.

At one stage the legend rivalled that of the Loch Ness Monster, striking terror into the hearts of farmers and tourists, and filling small children with dread.

Yet countless bounty hunts, safaris and expeditions - one conducted by Royal Marines - failed to pin it down. Sheep and farm animals continued to be mysteri-ously slaughtered across Exmoor. So one thing was probably on the minds of Sergeant Jeff Pearce and PC Chris Tucker when they were called to investigate a long-dead but fearsome creature washed up near Croyde Bay. Had they finally solved the riddle of the Beast?

A woman had reported spotting the remains of a creature 'the size of a calf with canine teeth'. She is said to have used the B-word.

The officers were duly scrambled. Once at the scene, a cursory glance revealed: Too big for a dog or domestic cat; too small for a pony. Wrong teeth for a cow. A seal? Not with those legs, it wasn't.

'It almost definitely looks like it could be a Beast of Exmoor,' said Sergeant Pearce, with admirable caution. 'It's only about five miles away to Exmoor by sea, it could easily have floated down.'

PC Tucker added: 'It's a good 5ft and it has black fur. It certainly looks quite beast-like with those teeth.'

So has the mystery been solved? Not quite. Samples sent for analysis revealed that the Beast of Croyde Bay was simply a grey seal. Decomposition meant its flippers had vanished to reveal bones that looked like they might have been limbs. Likewise, all that time in the water had given it a menacing snarl.

And the real Beast? There hasn't been a sighting for some time now, probably a week at least.

The legend lives on.

cerned that a specimen of potential scientific importance would be removed by the environmental health department, or chewed up by a badger, fox, or dog, and unilaterally decided that we should try our best to preserve it for posterity.

The beach is actually owned by a holiday camp, which in turn is owned by a firm based in Newcastle. We were unable to get all of anybody at the holiday camp, because it's the middle of winter, and nobody in their right minds would be on holiday in North Devon at this time of year. So, determined - as always - to be good law-abiding citizens, we telephoned the parent company. Nobody in the management department was available to speak to us, but we spoke to a delightful young lady called Gemma who told us that she was sure that nobody from a major holiday network would actually want the suppurating carcass remains of an unfortunately deceased pinniped, but nevertheless agreed to log my call and be witness to the fact that neither I, or anyone else at the CFZ has any intention of permanently depriving Parkdean holidays of part of a dead seal, and that if they want the skull and/or rear flipper back they only have to

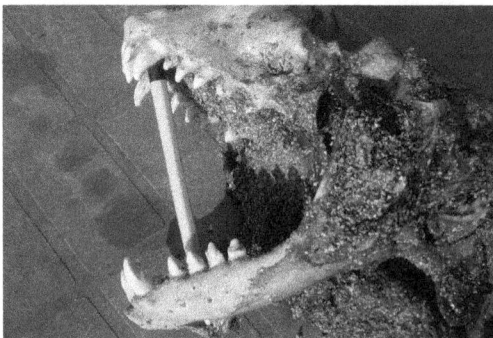

ask. Gemma thought that this was all terribly exciting, and rather amusing. She wished us luck, and our call ended.

After reading today's papers, CFZ Director Jon Downes (49) telephoned Braunton Police Station, and spoke to the Duty Officer, telling him the state of affairs, and explaining that the missing skull and flipper are at present in a bucket of formalin at the offices of the CFZ (the world's largest mystery animal research group). Disappointingly for his street cred, he was told that an immediate arrest was highly unlikely, and that the Police were merely happy that the cadaver was "in the hands of the professionals". On hearing this, Corinna Downes (52), Jon's wife, and administrator of the CFZ stopped making the placards reading "Free the CFZ Three" and went back to her normal activities.

Apart from the moments of intensely silly humour, the event does throw up some fairly serious issues.

Firstly. How did *The Sun* and also the local police find out that the skull had been taken? That part is simple. It was a local photo agency who were angry with us for taking the skull and wanted to get us into trouble. We were threatened with arrest twice (not by the police, I hasten to say, who could not have been nicer or more helpful) but by various people from the journalistic community.

Secondly. Since when has the testimony of some bloke on a beach been considered 'expert'.

Thirdly, we were criticised by some people within the cryptozoological community itself for making such a fuss about the skull. Admittedly, it was the usual suspects who are the ones who always seem to criticise the CFZA (and me in particular) at the slightest drop of a hat, but I feel that we should answer these accusations.

- The claim that the carcass of the `Beast of Exmoor` was washed up on Croyde Beach in January 2009 has entered the public domain. We are now in the position of being able to permanently disprove such claims. There are plenty of similar claims within the canon of Fortean literature which remain tantalisingly open, only because they weren't properly investigated at the time, and if they were proper records were not kept.

- There was a possibility, albeit a remote one that the carcass would have a *real* cryptozoological significance. One of the experts who first viewed of the photographs suggested that it looked like the skull of a sealion. There are seven species in the world, but with the exception of one species found on the coast of Argentina, all sealions are restricted to the Pacific Ocean. However, in the 1980s a Steller's sealion turned up on the Brisons - two tiny islets, a mile out to sea from the west coast of Cornwall. No one knew how it got there, and - as far as we know - it may still be there today. So, the mysterious cadaver of Croyde might just be only the second sealion ever to grace these shores.

The Brisons sealion

Zoologist Darren Naish, an old friend of the CFZ, who is probably most familiar to the world at large as the author of Tetrapod Zoology, an embarrassingly popular, and very informative blog which kicks everything we do into a cocked hat popularity wise, delivered his verdict.

"Hi Jon

Many, many thanks for uploading and linking to the pics - they are excellent, and I'm pleased to see that there are some with a scale. The *Daily Mail* photos made me think that it might be an otariid, but your photos demonstrate without doubt that it is a Grey seal after all. How do I know this?

-- In otariids, the nasals only extend to the anterior margin of the orbit: in phocids, the nasals extend much further posteriorly. Furthermore, otariids have bony lumps on the frontals (termed the supraorbital processes) that project over the orbits. Phocids lack these. The Croyd Beach skull has posteriorly extending nasals and no supraorbital processes. It is thus definitely a phocid (ʃ earless seal). I'm very surprised that some 'experts' said that it was not a phocid skull.

-- The nasal bones are unusually short in the specimen (compared to most phocids) and the nasal cavity is very deep, giving the anterior margin of the snout a steep, sawn-off look. Short nasals and a deep nasal cavity are both characteristic of Halichoerus, the grey seal. The skull definitely belongs to that taxon, case closed."

After the threats of our imminent arrest we had a long and extremely cordial conversation with Sgt Jeff Pearce of Braunton police station. He was very pleased to find out what we had done, and we promised him that we would keep him informed as to the progress of the inquiry.

However, he did give us another germ of information which both intrigues and disturbs us. Two national agencies were quoted in various newspapers as having said that the corpse was not that of a grey seal, but when I telephoned both agencies, they both insisted that they had

told everybody that it was a grey seal. I had assumed that this was the work of an over eager tabloid newspaper reporter, but Sgt Pearce, who came across as an eminently sensible and reliable bloke gave me the names of the people who he had spoken to at both of the aforementioned agencies, who had indeed told him that the corpse was not of a grey seal.

From a personal sense of inquisitiveness, and also from a professional point of view, I would love to know why these people, one of whom was quite a high-ranking expert, said what they did. However, it doesn't really matter, and I will not be pursuing the matter purely because I want to be (if I may misquote the Church of England, book of common prayer) in a state of love and charity with my professional neighbours, and these are people with whom it is quite likely that I shall have to work in the future.

So, the story is solved. It was a seal which may have had slightly abnormal nasal cavities. However, we are now on the position to be able to state this as incontrovertible fact. The CFZ went

out and got the skull, and will be keeping it in our museum because of the minor position which it will always hold in the history of cryptozoology. We have got a new friend in Sgt Pearce, and have been able to prove, as if any proof is necessary, that the tabloid press is not necessarily the best source of good scientific evidence.

And the final thing that came out of it was our now world-famous daily CFZ bloggo. Until early January we had been getting between 30-50 hits each day on our blog which was only updated occasionally, and was completely outstripped performancewise by sites such as Tetrapod Zoology and Cryptomundo.

After the surprising number of hits that we got on the blog with the stories that we posted about the Croyde cadaver (I always flatly refused to use the name `Beast of the Bay`) we decided to bite the bullet and launch a daily online magazine on the blog with an average of eight postings of original content each day. And so that's exactly what we did.

OBITUARY

If you want to read how John Michell was an inspirational figure to both the hippy and the Fortean movements, how he questioned the authorship of Shakespeare's plays, or how he co-wrote *Phenomena and* Living Wonders with Bob Rickard you will have to look elsewhere. Everyone else's obituaries of the man will be full of this stuff. I just want to use this space to mourn the John Michell who I knew, and of whom I was very fond.

Back when the CFZ first entered the international arena, we really were there under false pretences. I was strutting around the place claiming grandly that we were the largest cryptozoological organisation in Britain (we were, but only because there wasn't any realistic opposition), and whatever realistic credibility I had was only because of the kind endorsement I received from various members of the then Fortean establishment. And one could not get more part of the Fortean establishment than John Frederick Carden Michell. I met him at one of the early Unconventions and he was one of the most delightfully old-school Forteans that I have ever met.

The cliché `a scholar and a gentleman` come to mind whenever I think of John, because conversations with him would be ridiculously wide-ranging, but he would never talk down to anyone, even when - like me - they were a wannabe masquerading as being someone of importance.

We were friends for the next fifteen years, and the last time I saw him was at the Big Cat conference in Marston Trussell during the early spring of 2006. There he flirted roguishly with Corinna, skirted charmingly around the issue of whether he would ever write the book he had been promising CFZ Press for years, and drank a lot of my beer. He was delightful, and promised that he would come and do the Weird Weekend `one of these days`. Sadly, that's never going to happen now. **JD**

John Michell
(1933-2009)

AQUATIC MONSTERS LOG BOOK

BY OLL LEWIS

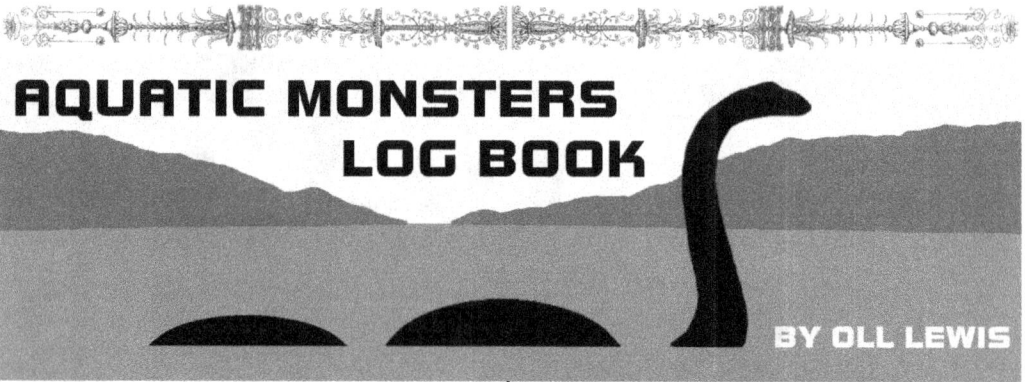

STATS AND SEA-SERPENTS

CFZ stalwarts Michael Woodley and Dr Darren Naish have, along with Dr Hugh Shanahan, published a paper in the journal 'Historical Biology' that predicts there could be three species of pinniped yet to be discovered.

The conclusion was reached after eyewitness sightings of unknown sea animals were examined in detail and applied to two different statistical models. One of the statistical models predicted that there could be as many as fifteen undiscovered species of pinniped. But Woodley considers this may be an over-estimation because the statistical model cannot take into account the fact that - due to their close ties with the coast and the rather vocal nature of pinnipeds - it is more likely that pinniped species will be encountered by man than other sea mammals, such as whales which live most of their lives in the open oceans.

It is postulated that among the unknown species of pinniped there may be examples of the 'merhorse' (a 4-30 metres long supposed deep water animal, a carcass of which was allegedly found in a whale's stomach in 1937), the long necked sea-serpent (a creature with a similar silhouette to a plesiosaur) and the Tizheruk (a pinniped like sea-monster with a long body and snake-like head described by witnesses and myths from the Pacific Arctic).

"We consider that if such creatures as the mer-

AQUATIC MONSTERS STUDY GROUP

www.cfz.org.uk

horse and long-necked sea-serpent exist, they must be extremely rare. They must also dwell in remote and seldom visited regions of the oceans," Woodley cautions.

STORSJÖN NORSEMAN

Last issue I wrote about the monster of Storsjön Lake (aka Storsjöodjuret) in Sweden having been allegedly captured on film and of the continued

resolve of the group responsible to continue filming the lake should the creature show up again. The team has not captured anything unusual on film since the blurry video last August, but they have fallen foul of the local authorities. After the film was released, Jämtland County Administrative Board looked into whether the team had applied for a permit to film and publish footage from the lake's shores, which appeared on their website. They had not, so the council issued fines.

"*When we found out about the camera surveillance undertaken without a permit we immediately conducted a review. The association has since been encouraged to apply for a permit,*" said Åsa Johansson at Jämtland County Administrative Board.

After a review of the case, the council agreed a permit for four underwater cameras but not for surveillance cameras to be placed above the waterline, which was ruled to be in breach of regulations.

This might well be the first case of a large-scale lake monster hunt being hampered by NIMBY-ism. I could understand cameras being banned if

they were positioned to look into people's bedrooms or public toilets, but the sort of people who object to fixed surveillance cameras positioned to look out at a lake are either the sort of idiots who object to wind turbines for aesthetic reasons or perhaps up to no good.

DON'T NESS UP THE BID

Loch Ness has applied to be considered for UN-ECO world heritage site status, which would place the loch on a par with such wonders as the Great Barrier Reef, the Taj Mahal, the Grand Canyon, the Great Wall of China and Big Pit, Blaenavon. Loch Ness, like many places in Britain, is a place of outstanding natural beauty and certainly deserves to be recognised on an international level, but in order to become a world heritage sight, a sight must have a unique quality that really sets it apart from other similar places around the world. It is because of this criteria that even Snowdonia and Ben-Nevis would be unlikely to qualify as world heritage sites whereas the city of Liverpool (I've lived there, great people, but a lot of derelict buildings) and the aforementioned Blaenavon qualify because of their history.

This shouldn't be a problem at all for Loch Ness then, because it has its most famous resident; the Loch Ness Monster.

Even if Nessie doesn't exist the social history surrounding what is, without a doubt, the world's most famous cryptid would be a certain deal clincher in this matter. When I was a student I was having a conversation with a flatmate from Nepal about unusual animals, hoping to coax him into telling me some locally known tales about the

yeti that had not made it to the west, turned out he had never even heard of the yeti, but had heard of Nessie!

All this makes the fact that the people responsible for the bid are proposing not to mention Nessie in it, for fear of the bid not being taken seriously, all the more bizarre. You can imagine several come-dic scenes when the UN inspector turns up to as-sess the site's world heritage claim with local officials trying to create a diversion as they go past places like the Loch Ness Exhibition Centre lest the investigators ask what all the wee statues of plesiosaurs are about.

All joking aside I can't honestly see Loch Ness attaining world heritage status if the monster is omitted from the bid, which will make it a huge wasted opportunity and a waste of public money.

SUPER SALMON

In October last year the largest Chinook salmon ever found in California washed up dead. The salmon measured over 51 inches and is believed to have died from natural causes before washing up on the banks of Lower Battle Creek, near Red Bluff, California.

Biologist, Doug Killam, said that the fish could have weighed more than the 88 lbs salmon caught in the Sacramento River, which is the largest re-corded Chinook salmon in California. The heavi-est Chinook salmon ever caught was 126 lbs in Petersburg, Alaska, but most of the salmon biolo-gists have weighed in California are between 20 and 30 lbs.

STINGRAY, STINGRAY!

© BNPS.CO.UK

The world's largest fish ever caught on a like and rod has been landed in Thailand.

The freshwater giant stingray is estimated to be between 550 to 990 lbs (250-450 kg) in weight. The stingray's body measured 6.6 feet (2 metres) wide by 6.9 feet (2.1) metres long. The tail was missing. Had the tail been present, the ray's total length would have been between 14.8 and 16.4 feet (4.5 and 5 metres), estimated University of Nevada Biologist Zeb Hogan.

The stingray was caught by Ian Welch, a profes-sional fisherman, biologist and columnist for the magazine Angler's Mail, who was visiting Thai-land to help with a stingray tagging programme when he landed the monster in the Maeklong river. The 45-year-old said he was nearly pulled over the side of the boat when the specimen took his bait.

"It dragged me across the boat and would have pulled me in had my colleague not grabbed my trousers – it was like the whole earth had just moved. I knew it was going to a big one," Welch related.

"It buried itself on the bottom and the main fight was trying to get it off the floor. I tried with every ounce of power but it just would not budge. After half an hour my arms began shaking and after an hour my legs went. Another 30 minutes went by and then I put a glove on and physically pulled the line with gritted teeth and somehow I found the reserves to shift the fish." Once the shed sized stingray was off the riverbed it became much easier to lift to the surface, but it still needed 13 people to land.

"As soon as we saw it there was just silence because everyone was just in awe of this thing," he said. *"That line from the film Jaws came to mind about needing a bigger boat because we had to get it to the shore to tag it."*

"It took 13 people to lift it into a large paddling pool we had set up in order to tag it and take DNA samples."

"I was absolutely exhausted afterwards and did very little for the rest of the day and just had a cold beer. As a life-long angler and a biologist it is great that my two passions have come together and culminated in something I could only have dreamed of."

Hogan, the biologist who estimated the creatures size and landed another large stingray in March 2008, believes that catches like these provide evidence that the giant stingray may not be as rare as previously thought:

"The Thai populations were once considered critically endangered, although with the discovery of new populations the stingray's abundance appears higher than previously believed."

Little is known about the lifestyle and ecology of the freshwater giant stingray, which was first described only in 1989, and it is listed as vulnerable by the IUCN.

OAR-SOME FINDS

You know how it is, you wait ages for something then a few come along at once. I've just covered record breaking freshwater giant stingrays, two of which have been caught in the last year and now three oarfish (*Regalecus glesne*) have washed up on the English coast in just over two months. This is all the more impressive when you consider only four oarfish have been found washed up on the British coast in the last 20 years.

The first two oarfish were both found near Amble in Cumbria on Boxing Day and in early February by fisherman Michael Bould and his wife. The second fish was apparently alive when found by Mrs Bould:

"She photo'd it and with her son pushed it back into the water. When I came ashore from fishing on Tuesday lunchtime I took the dog for a walk with the outside chance of getting a glimpse. After walking to tide edge for about 1.5 miles south of the find site, and about to give up I saw a flash of silver half way up the beach and unfortunately the fish was there having died and washed up overnight." Michael Bould related to the 'Boulmer Birder Blog'

It is not certain that the second dead oarfish is the same one Mrs Bould re-floated and given the regularity they have been turning up in the area it is possible that the second dead fish might have been the third fish to be washed ashore. Regardless, the third dead oarfish was found on a Tynemouth beach, North Tyneside in near perfect condition. The Tynemouth oarfish is 3 m long and was recovered by the Blue Reef aquarium who will be performing an autopsy on the fish.

Zahra d'Aronville, the curator of Blue Reef said of the Oarfish: "Very little is known about the life cycle of these truly magnificent creatures and it's a mystery why two of them should have washed up on our coastlines so close together."

Because at least three oarfish, have washed up in a short space of time one possibility is that there is a congregation of oarfish off the North East of England coast meeting up to feed or breed.

Oarfish are known to grow to 8 m long but there are some reports of oarfish growing as long as 16 m and it has been postulated that some sea serpent sightings may have been sightings of oarfish.

EEL GOES ROGUE

An eel longer than 1m long has recently been attacking swimmers in Australia. The eel has been attacking swimmers in Gwangoorool rock pool, Binna, Burra on Australia's Gold Coast. The eel has become so violent that local authorities have had to put up signs warning swimmers that eels inhabiting the water have been known to attack.

Once of the eel's recent victims, student Carolina

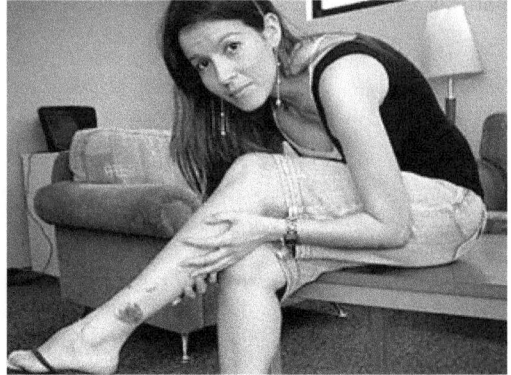

Franco, needed tetanus shots after she was bitten twice on the leg.

"*It really scared me,*" she said.

"*We saw the eel swimming around when we got there, but we didn't think eels would attack people so we weren't worried.*"

Rangers for the Environmental Protection Agency confirmed that they would not be attempting to move the eel.

"*Eels are part of the natural environment at Binna Burra and we need to remember we are entering their habitat. The EPA won't be moving them out of their pool*" said a spokesman.

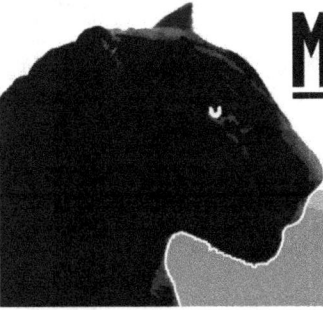

MYSTERY CATS DIARY

MYSTERY CATS OF MICHIGAN

by Raven Meindel,
CFZ-Michigan Representative

Note from Nick Redfern:

As readers of the CFZ's blog *Cryptozoology Online: Still on the Track* may be aware, I recently set up a new blog – titled *Crypto Squad USA*. And what, you may ask, is *Crypto Squad USA*?

Well, that's a very good question! No, *Crypto Squad USA* is not a bunch of super-heroes - but they are pretty close! When I moved to the US to live in 2001, Jon asked me to take on the task of opening, running, and coordinating a US office of the CFZ - which I was pleased to do.

But, as CFZ-USA began to grow, it quickly became apparent that regional representatives were needed to cope with the huge amount of field-work and investigations that were required. So, that's precisely what *Crypto Squad USA* is: namely, a dedicated band of people seeking out all-things strange and monstrous on behalf of the CFZ-USA Office, and making the data available to one and all.

If you're interested in becoming a regional investigator for the US Office of the CFZ, let me know! We're always looking to get new people involved. You can contact me at nick_redfernæsbcglobal.net or at the *Crypto Squad USA* blog address: http://squadcrypto.blogspot.com/

And with all that now said, I am very pleased to be able to present you with an insightful article from a very good friend of mine, Raven Meindel, who has a deep interest in cryptozoology – and particularly in her home-state of Michigan, for which she is now the CFZ-USA representative.

Ask most of Michigan's Department of Natural Resources (DNR) officers about cougars – or, for that matter, about any big-cats here - and you're bound to get an emphatic negative response. But, talk to some of the local folks in our great state and you'll find that big-cat encounters are not only a reality, but the beasts seem to be thriving and flourishing in several regions of Michigan.

The fact that they appear to be quite at home in this environment isn't so surprising once you consider the fact that, between 2 and 10 million years ago, *Puma concolor* (the cougar) roamed freely all across the continent, and is believed to be responsible for pushing his cousin, the jaguar, into its now native South America.

Paleontologists have established a common ancestry of all big-cats which began approximately 30 million years ago in central Asia - meaning the cats crossed the Bering land-bridge to establish their presence on the North American continent.

A felid that reigned unchallenged for a period of time was *Panthera leo atrox* (a.k.a. the American lion or cave lion). In appearance, it was much like the African lion, but quite a bit larger, weighing in at an average of 500 pounds or more. It became extinct at the end of the Pleistocene era......or did it?

Michigan has certainly had its share of encounters with felids of all types. There have already been seven big-cat encounters in the region this year, with two of those being of the ever-elusive mystery black cat variety.

Throughout 2008, there were many sightings of large black felids, with the majority of sightings occurring in the areas of Oakland County, Eaton County, Kent County, and in my very own Washtenaw County. One large black cat sighting was in November 2008, literally just down the road from where I reside.

A well known encounter that became a bit of a sensation occurred on April 20, 1987 in a little place called Wixom city, Michigan, which is a part of Oakland County, and which dates back to the 1830's.

It was a warm spring evening around 10:30, and Jimmy Trick had his windows open to take in some of the fresh night air coming in off of Loon Lake. Earlier in the evening, Shiloh, his black Labrador, was barking and carrying on at the window; but this was not an uncommon occurrence, given the breed of dog, and the fact that Loon Lake was home to a variety of wildlife.

Jimmy didn't give it a second thought at the time.

However, the silence of the night was suddenly broken when Jimmy heard something of a considerable size splashing around down in the canal in front of his home. The canal itself was roughly 20 feet in width at the time (today, it is mostly dried up), and there were reeds, cattails and spongy ground on the other side of it.

Jimmy wasn't alarmed at first: just curious. He turned on the floodlight, grabbed his flashlight, and began making his way down to see what it was that was making such a ruckus in the water.

As he approached, the noise stopped abruptly - almost purposefully. Whatever was splashing in the canal had now turned its attention on Jimmy; he flashed a beam of light into the reeds, and caught a glimpse of two big green eyes staring back at him. The creature was fairly large, said Jimmy, who estimated it at approximately five-to-six feet in length, and black as the night that now enveloped both of them.

As Jimmy started to move, the creature also began to move. It walked parallel to him on the other side of the canal - as if stalking him. It was at this moment that it dawned on Jimmy that the creature could easily clear the twenty-feet of water between them, and he began back-stepping up the embankment to his house.

When he reached the front door, Jimmy quickly darted inside and phoned the local police. When they arrived, however, they had a typical air of skepticism about them, but took down the report in the usual way.

In an odd twist of events, the officer who had taken the report had an encounter of his own, when the creature ended up in full view in his backyard the following weekend. A couple of other locals were known to have caught sight of the cat as well. Though no more activity has taken place in the area since, the sightings were not forgotten, and hats were even made to commemorate the occasion.

As Michigan-based big-cat encounters continue in the coming years, as they surely will, we will continue to research and learn more about how they've come to be here and why. And as we gain knowledge and understanding, hopefully it will mean a willingness to live in peace with them, and alleviate the fear that often comes with the arrival of such a predator.

Was he always here, and just able to remain elusive? Or was there some artificial means or intervention - intentional or otherwise? These are the questions that I and other fellow researchers hope to soon answer. Thanks again to Jimmy Trick for sharing his amazing story with me.

CFZ Representative for Michigan
734-663-6526
http://ravensmysterioushaven.blogspot.com/

COUGARS IN ILLINOIS

Derek Grebner
CFZ Illinois Representative

Despite being officially listed as having been extirpated in the 1860s, sighting of mountain lions are surprisingly prolific in Illinois, and it sometimes seems that "everyone and their brother, and their dog" have seen mountain lions in Illinois. Even I, in my youth, may have found the tracks of a cougar. I knew they might have been because, they did not have claws in the track, and coyote or dog track have claws in their pugmarks. As I am sure you all know, cats have retracting claws and so their claws do not show up in their pugmarks.

Since 2001, two mountain lions have been shot and killed in the state.

The idea is very provocative, but some from the newspapers outdoors column and the DNR, feel that the mountain lions being seen and killed in Illinois are from the black hills population. It is not impossible that the animals could travel that far. Considering the amount of sightings over the years that been reported, and those which must have gone unreported, that I have come across at campfires in my life, it just do not sit well with me that they are all travelling cats.

I am in direct contact with the newspaper and many outdoorsmen So, I will know as soon as anything new comes to light.

NICK REDFERN'S
Letters from America

Well, there are a couple of things to report on in this issue of *A&M*.

First: things are progressing very well over at the new CFZ blog which is dedicated to the work of the Centre's US Office (titled *Crypto Squad USA*), and which can be found at http://squadcrypto.blogspot.com

Amongst the latest entries, you can find a number of intriguing and newsworthy blog-posts from the many and various US-based CFZ representatives. There is, for example, Melissa Miller's article on Oklahoma's big-cats, in which she states in part:

'Big cats are not new to Oklahoma. Mountain lions have been documented in the state since 1852 and listed as a game species since 1957. The Oklahoma Wildlife Department describes the cats this way: "Its tail is more than half the length of the body, it has black tips on the tail and ears, and is primarily tan in color. The size of these animals varies by sex. Males average seven feet long (from nose to the tip of its tail) and weigh around 140 pounds, while females average six feet in length with a body weight around 95 pounds." The Oklahoma Wildlife Department also states that the best place to see a cat is extreme western Oklahoma; and yet, some of the most dramatic sightings have taking place in extreme eastern Oklahoma.'

Our Oregon-CFZ rep Regan Lee has a thought-provoking new post, too. And, as with Melissa's post, here's a taster of Regan's:

'National and smaller, independent restaurants often promote their food offerings using the very thing they kill and sell to us to eat. On the surface, it's presented as humour, nothing more. The chicken, fish, cow or pig joyfully entices us to partake of him. Just below the surface is the idea of the mascot representing another realm where the spirit of the sacrificed animals calls to us, bewitching us to eat its own kind. Below that, however, is a symbolic exorcising of guilt; by using the very animals we're killing and eating as a happy and enthusiastic ambassador, we don't have to deal with our responsibility in the process.'

Remember that the next time you munch on the cholesterol-burger of your choice!

Meanwhile, the CFZ's Michigan-based representative - my good friend Raven - has been doing a tremendous amount of work: writing articles, doing on-site investigations, establishing new links with people in the field, as well as setting up websites and blogs.

Does the woman ever sleep?! Probably not!

Anyway, Raven has an interesting new post titled *Knock Knock Bridge*, which is a study of a wealth of weirdness at a particular bridge not far from where she lives.

I particularly like stories like this, as I've done a lot of investigations myself where high-strangeness has occurred at bridges - most famously from my own perspective at Bridge 39 on Britain's Shropshire Union Canal, where a mysterious, spectral ape known as the Man-Monkey was seen in 1879, and whose exploits are chronicled in my *Man-Monkey* book.

And with that said, here's an extract from Raven's post:

'Today Jessica [Raven's daughter] and I took a ride out to the Canton area to do a pre-investigation study of the Denton Road bridge and the surrounding area. There is so much history that lies just beneath the surface of that entire area that we'll have to do many investigations to cover even a portion it.

'As we drove through where the bridge is located, an eerie ambiance fell over the van. It was like we had just slipped into another dimension of time. With the exception of the newly built condos that now dot the land heavily, and the occasional jogger or bicycle, there is a stillness to the place that can only be described as otherworldly.

'We met a gentleman named Don who has lived there since 1963 and was not only able to give us a first-hand account of the folklore, but also had a wealth of knowledge regarding the development of the area as well.

'Don told us that the bridge was called "Knock Knock Bridge" by all the local children when he was growing up, because, as legend has it, if you knocked three times and waited, something spooky was bound to happen.

From ghost lights to shadowy dark figures chasing cars, there have always been strange stories passed down from one generation to the next.'

And finally, this is more of a news-story combined with a plea for data.

I referred earlier in this issue's column to my *Man-Monkey* book, which is a study of a weird, spectral-ape-style beast seen in 1879 (and, according to some, to the present day, too) in the woods that surround England's old Shropshire Union Canal, and amid the small villages and hamlets of Shropshire and Staffordshire that straddle the canal.

Well, the good news is that Jon has agreed to publish a new book from me to be titled *Wildman! On the Trail of Britain's Bigfoot and Other Man-Beasts*, which – as its name strongly suggests, of course - will be a full-length study of reports of Bigfoot and other mysterious ape-like and hairy, upright creatures seen in the British Isles.

Far broader in range, time-frame, scope and con-

tent than *Man-Monkey*, *Wildman!* is likely to be a (literally) huge book that encompasses everything from the aforementioned Man-Monkey to the Beast of Bolam, from the Big Grey Man of Ben Macdhui to the Shug-Monkey, from the Ghost Ape of Marwood to the Beast of Cannock Chase, from the tale of the Hairy Hands to that of the Hexham Heads (yep: werewolves will be in there, too), and much, much more, too.

Of course, the British Bigfoot is far stranger (and that's putting things mildly!) than its many-and-varied overseas cousins: indeed, it has much more of an air of the "Zooform" about it than it does of anything of a flesh-and-blood nature.

And so: some of the theories that will be presented within the pages of the book are likely to please some, intrigue others and downright irritate a few more. But, so be it.

There is no way that a colony of large ape-like beasts roams the British Isles. Yet, people are seeing something.

And it is this conundrum (along with a multiplicity of reports spanning more than 1,000 years) that the book firmly addresses via a variety of theories, thoughts, observations and ideas.

So, back to my plea: if anyone reading this has had an experience with the so-called British Bigfoot (or something similar in Britain), or wishes to have their views included in the book via an interview, or a statement in their own words, do let me know.

All data, sighting-reports, and commentary on this highly-controversial issue will be greatly appreciated.

You can contact me at:
nick_redfern@sbcglobal.net

Rumours that the front-cover of *Wildman!* will feature a photo of Jon racing around Woolsery in a crazed fashion by the light of a full winter's moon just recently can be neither fully confirmed nor denied at this time...

BIGFOOT NOTEBOOK
Paul Vella

For readers, *Animals & Men* seems like the sort of publication that turns up on the doorstep once in a blue moon, but for me, it seems that no sooner have I put a BHM Roundup to bed and Jon Downes is on the phone telling me to get my act together and write another article. *"What do you mean you are overworked and haven't slept in two weeks?"* he says, *"Animals & Men is surely the most important thing in your life dear boy"*.

(You mean it isn't the most important thing in your life? I am seriously disappointed with you, Vella JD)

I mumble something about having it ready by the end of the week, wondering how long I can go without sleep *"and how is the book you promised me coming along?"* asks Downes, *"on target"* I replied, and muttered some un-printable expletive under my breath before adding *"you'll have it by the end of the year"*. I hung up the phone and went to find some caffeine delighted that he forgot to remind me about the blog I was supposed to be contributing to....

And so another BHM Roundup was started, and to be perfectly honest, I wasn't entirely sure what to include in this edition, since frankly, very little has happened since the last roundup, you see the world of bigfoot seems to go quite during the winter months when it becomes difficult for researchers to get into the woods. That might sound like something of a cop-out, but you have to understand that many forest roads are closed by the parks service for months at a time, restricting access to area hundreds of thousands of acres over winter. It occurred to me that these could be wonderful opportunities for wildlife to go about their daily business without interference from humans. In the case of bigfoot, the creatures (with the usual 'if the really exist' caveat) would

be free from prying eyes for a significant portion of the year.

Back in 2001 when Britain was in the grip of a foot, you will recall that the Forestry Commission closed all the public footpaths. Behind my Northamptonshire house is a glorious ancient forest, which is teeming with wildlife that as with most forests, manages to stay completely hidden from view. In all the years I had walked in these woods, I had seen just two deer. For several months during the summer of 2001, I could not take my dog for a walk among the trees. I kept checking the Forestry Commission website for news of the ban, and one fine morning, the website declared that this forest had re-opened. I grabbed my Jack Russell's lead and told him we were going to the forest – he barked with delight, or at least I think he did, it was the same bark he used when he chased tail. Anyway, the point of this story is that on that wonderful late summer's morning, the forest was teeming with deer – it was as if they had become used to not having humans around, and were quite literally out in the open. On at least half a dozen occasions we came face to face with a deer that was just standing, bold as brass on one of the trails.

With that in mind, I can't help but wonder whether we would see more evidence of bigfoot if the US researchers could get deep into the forests during the winter months and be in places where they quite literally aren't supposed to be.

The Minnesota Iceman

I was unable to make last year's Weird Weekend for personal reasons, but before I forget to mention it, I will be presenting a talk at this year's Weird Weekend on the Minnesota Iceman. As usual, I will be bringing alleged footprint casts and all manner of documents to be put on display.

I look forward to seeing you there.

Georgia Hoax

Just when you thought the Georgia 'Bigfoot Body' Hoax had been and gone, the two main protagonists reared their respective ugly heads again recently. In a January podcast hosted by New York researchers Steve Kulls, Matt Whitton and Rick Dyer stated that veteran researcher Tom Biscardi was in on the hoax from the beginning. It is always difficult to believe people who hoax, but there may be some merit to this, since we know that Tom Biscardi first visited the 'Georgia Boys' on 1st August 2008 and claimed to have seen the body in the freezer. Just one problem, and that is that the freezer wasn't purchased (according to the invoice) until 8th August 2008, the day Biscardi made his second visit to Georgia. Something smells, and it isn't just the fake corpse.

Sensing that something wasn't quite right, the other Bigfoot demanded to see the body.

San Francisco Chronicle

On 24th April this year, the struggling *San Francisco Chronicle* published an article 'Five Places: Walk in the shoes of bigfoot'. The article listed a number of places of interest in California, including Happy Camp, The San Diego Museum of Man, the Bigfoot Lodge in Los Angeles, Willow Creek and the Bigfoot Discovery Museum in Felton.

Of these places, the San Diego Museum intrigued me, apparently it has a life-size model of *Gigantopithecus blacki*, but I would definitely recommend a visit to Willow Creek, it is a very charming small town in Northern California, with one of the finest collections of bigfoot artefacts and museum pieces I have ever seen. I have yet to visit Mike Rugg's Bigfoot Discovery Museum in Felton, but friends who have visited tell me that it is well worth the trip if you are ever in that part of California.

The original article can be found via this link: http://tinyurl.com/dhrask

Russian Search for the yeti of Shoria Mountain

The search for the yeti of Shoria Mountain in Russia is set to continue as the snows melt and the weather warms in Siberia. The expedition will commence in June 2009 and will be headed by Valery Kimiev and Professor Nikolai Skalon, the head of the department of Zoology at the Kemerovo State University. Local's claim to have seen the creature in the area surrounding the Azass Caves on many occasions. One local, hunter Michael Kiskarov will be joining the expedition. Kiskarov claims to have seen a Yeti on more than one occasion.

The local head of government Vladimir Tashtagol has received fourteen written reports by residents of remote villages around the area of the Azass caves and the nearby Mrassu river claiming to have seen these creatures in recent years. The Cryptids are described as being two meters tall, with reddish black hair that resembles that of a brown bear. The Azass cave is several kilometres deep and it is believed the yeti or yeti's dwell deep in its interior. Some believe that the creature is a lone Neanderthal survivor or the sightings have been of one of a number the species, part of a small colony that represents the world's last Neanderthals.

Abominable Snowmen Online

The full text of Ivan T. Sanderson's 1961 book *'Abominable Snowmen, Legend Come to Life'* is now available to read online at www.sacred-texts.com/lcr/abs/

Northern Territory man claims Bigfoot-style Yowie killed his dog

The following article was published in the *Northern Territory News* on April 21st 2009, and was followed a day later by a second article where other researchers claimed that the Yowie was being unfairly blamed for the death of this dog. Curiously, I have read numerous reports over the years of North American bigfoot killing dogs.

IT'S feared Yowies could be on the loose in Darwin's rural area.

A Territory Yowie researcher believes the Bigfoot-like beast could be responsible for the recent death of a dog south of Darwin, The Northern Territory News *reports.*

The dog's owners believed their seven-month-old puppy, which had its head ripped from its body, was mauled to death by dingoes.

But Andrew McGinn, who has been researching Yowies in the Top End for more than a decade, said it was possible the hairy ape-type beast was responsible for the attack.

"The way the guy's dog was killed was typical of a Yowie," he said.

"I know it sounds fanciful but over the past 100 years, dogs get killed or decapitated and people report feeling watched, having goats stolen or seeing some tall hairy thing in the days beforehand."

In the late 1990s there were several reports of Yowie sightings around Acacia Hills.

Sightings

The trouble with sighting reports is that there is usually no way to verify them, and so I present you with a few snippets from reports that have come to light recently. I have removed details from some that might identify the witness or location. I have not corrected spelling or grammar.

This report is from **1985**:

The boys were fishing at the pond, which was just up the hill by our house. They had started a fire when "something" from on top of the high wall started throwing rocks, etc. They were not aimed at the boys.. "It" just wanted the fire out. The boys come running down to the house-so upset- over this "big thing" that was throwing rocks. I sat them down and asked them if it could have been the old man that lives up the road, I asked many questions hoping to come up with some kind of answer beside it wasn't human. I drove them up to the pond to get their gear.. at that time I felt weird, all that I heard was the bushes moving. I could smell this odor that was like nothing I had ever smelled. The boys at that time knew what they saw, but have somewhat dismissed it because of all the teasing.

I sometimes wonder just how common footprint finds are. I suspect that most are misidentified bear prints. Bears sometimes walk overlaying their hind feet where their front feet have just been, with the effect of elongating the footprint. Regardless, here is a typical report from the Alliance of Independent Bigfoot Researchers:

*In 1997 Me and my prospecting buddy where walking into a good panning location on the **** rd. We where the first ones in that year. The roads still had a foot or two of snow in places. The roads where still washed out then from the 1996 floods. About two miles in past the road block at the washout. A big set of tracks came down off of an extremely steep hillside on the uphill side of the road and crossed to the far side of the road and continued along the road for some distance. I assumed it was just an elk or something and ignored it. My buddy Frank at some point stopped to take a leak off the edge of the road. At wich point hey got a strange look on his face and said " Hey man , you should come look at this!". So I go over and he points at the track. It is a very large (bigger than my size 10 1/2 boot) barefoot footprint. Looked just like a genuine Bigfoot print right outta the movies. After trippin on it for a couple minutes we continued along our way. The tracks went a bit farther up the road and then dropped off over an extremely steep hillside down toward yellowjacket creek.*

The amount of difficulty and effort involved in faking a set of tracks like that is really unthink-

able. I don't think anyone would bother to take the time to do that there. There are no roads above or below the 29 at that point and it was in 2 feet of snow. Mine and Franks boot prints where the only human tracks in the area. With that in mind, well, I guess that there might really be some kind of bigfeet out there.

This report was published by the Texas Bigfoot Research Conservancy (www.texasbigfoot.com)

I scouted the area the day before and hung my ole' man climbing tree stand on a tree for the following morning's hunt. The only tree I could use was a small oak with a main fork that was about 12 feet off the ground and covered with poison oak. That means I could only climb up to that fork, and not my usual 20 or 25 feet which I usually like to hunt from. I'm not allergic to poison oak, so that wasn't going to bother me. I arrived early (4:45 A.M.) parked and got dressed. I've learned that to be a successful hunter, you have to be scent free. I dressed in my realtree leafy suit and lacrosse rubber boots, sprayed down with scentkiller and dowsed my boots with persimmon scent and walked to my stand (5:00 A.M.). The Moon was full and I could see with the use of my Streamlight (green light) pen light.

With the morning dew, I made no noise getting to my stand and without scaring game off with a large flashlight. I made it to my stand and found that the top portion was twisted around to the back of the tree and the bottom was moved to the base of the tree. Why did someone do this and not steal it? (5:20 A.M)

I was up the tree and had my bow in my lap with my arrow pointing away from me with it resting on my stand's gun rest. I pulled my earth scent wafers out and attached them to the outside of my back pack.

I was facing south with two big oak trees about 25 yards away. The clouds were moving fast and with the full moon, I could somewhat see in the dark.

I was very still and had everything camouflaged except my muzzy broadhead, on the end of my arrow. It was about 5:30 A.M. when I heard something walk up from behind me.

I turned slowly to see if I could see it. It was big and black and at first thought it was a couple hogs. They were too big to be hogs and I wondered whose cows were loose in the refuge. I was wanting them to hurry up and leave so they wouldn't scare the deer when something brushed up against my leg. I didn't hear anything climb up the tree with me so I kept still. I was looking forward when I saw a big black hand reach up and grab my muzzy broadhead. The hand was a shiny black and the fingers were huge! The palms were also black like a gorilla. When it grabbed my broadhead the razor blades cut him and it yelled like a bull! It was louder than a car horn and it was at my feet. It scared the crap out of me so I started yelling back.

And cussing! It ran off braking trees and tearing down everything in its path. I sat there 15 minutes until it got daylight and left. I believe that I scared him/her as much as it scared me.

And of course, one final historical sighting to round off this roundup...

The Van Wert Daily Bulletin.
VAN WERT, OHIO, SATURDAY, OCTOBER 28, 1905.

BRITISH COLUMBIA MOWGLIS

Tribe of Wild Men Roaming Woods and Frightening People.

James Johnson, a rancher living near Cornox, seven miles from Cumberland, B. C., reports several Mowglis, or wild men, who have been seen in that neighborhood by ranchers, says a Nanaimo (B. C.) corr-spondent of the San Francisco Call. Johnson asserts that they were performing what seemed to be a sort of "sun dance" on the sand. One of them caught a glimpse of Johnson, who was viewing the proceedings from behind a big log. The Mowglis disappeared as if by magic into a big cave.

Thomas Kincaid, a rancher living near French creek, while bicycling from Cumberland, also reports seeing a Mowgli, whom he describes as a powerfully built man, more than six feet in height and covered with long black hair. The wild man upon seeing Kincaid uttered a shriek and disappeared into the woods. Upon arriving home Kincaid wrote Government Agent Bray of Nanaimo, inquiring if it would be lawful to shoot the Mowgli, as he was terrorizing that vicinity.

The government agent replied that there was no law permitting such an act. It is reported that on a recent hunting expedition up the Quailicum river an Indian saw a Mowgli and, mistaking him for a bear, shot at and wounded him. During the past month no less than eleven persons coming to Nanaimo from Cumberland have seen the wild men. Parties have been organized, and every effort is being made to capture the Mowglis.

ON COVERT COCKROACHES, SECRET SNAKEHEADS AND PUZZLING PISCES

Max Blake

Every cryptozoologist has his or her own working definition of a cryptid. For some, it must be something of the megafauna variety, the bigger, scarier and more impossible the better. The company of these people I avoid. My definition is of anything that is not scientifically described. Why? Simple: it is hidden (*cryptos*) from science. This includes large apes from around the world, odd fish lurking at the bottom of our largest rivers, and the innumerable species of arthropods, molluscs, annelid worms, plus a whole assortment of animals from smaller phyla.

Small animals speciate rapidly: they are more prone to geographical isolation than bigger animals (which can, for instance, fly or walk over hills impassable to a millipede for instance) are, making it often hard to describe them as they may either change over time to split into two separate species, or have so many varied sub-species that it is hard to separate out if there is a different species within the group. An estimate for the total number of beetles alone (currently at 370,000 species, and rising) stretches into the multiples of millions, a very conservative guess being 2 million. Considering this, it is logical to see that we have barely made a scratch on the surface of what is out there.

A pastime even older than classification is that of keeping exotic animals. Recently, there has been a surge in the demand for new animals in the trade to keep it exciting (well, more than normal) and the collectors have gone a little overboard in their species collection.

Normally when a new species is found, it is dropped in some strong alcohol and sent to a museum. A while later when someone has found time to go through the keys to describe it, it gets a scientific name. Then, the collectors go out, grab a load, and send them off to wholesalers, and then to hobbyists. Most of the animals end up in places like Germany, the Czech Republic, North America, the UK and various other high-income countries. They are bred, sold, and become part of the hobby. Now, collectors seem to be skipping the museum part, and send them straight off to wholesalers and hobbyists.

The consequence of this is that we end up with loads of animals which are not scientifically described being kept by hobbyists. Often little is known about where the original animals came from, so there is little hope of getting out and finding more examples.

Quickly, we will now look at groups of new species that are turning up in the hobby that have piqued my interest.

My favourite groups of animals to keep include cockroaches, beetles, cichlids, cyprinids, snakeheads and scorpions, and it is in these groups that we often see new species popping up in the pet trade. We will ignore scorpions and beetles as although new species do turn up, they are nowhere near as common as in the other groups. Lets start with the fish.

As Jon has just found out, by buying a copy for CFZ HQ, the *Baensch Aquarium Atlas* photo guide is an amazing book. Over 4,000 species are shown with 4,600 photos, a 30,000 entry index and brief information on each fish featured.

Want more information? Have a look in the appropriate Aquarium Atlas between vols. 1-5 to find it (not hard for me, I have all of them). It really is marvellous. I did a funny little experi-

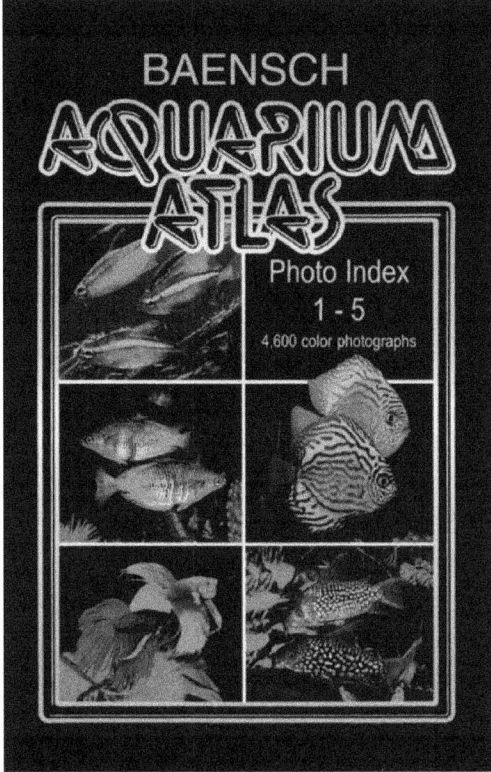

ment yesterday to see a representative sample of the number of undescribed fish contained within its walls.

On one page of cichlids, we had, out of 10 shown, 6 were undescribed. To show this is not a fluke, I turned the page. Yet more undescribed species. I turned again, and so the pattern continued. Ok, so 6/10 is a bit higher than average; but I would be willing to bet conservatively that about 1/5 of all the species contained within are undescribed, a rough total of at least 800 fish! That is a hell of a lot! Needless to say, not all of these will end up being (if they are actually described) new species, but may be new sub-species, colour forms or regional variations. Of course, few of these are available in your average fish store, but go to the right one, and you can find lots of tasty new species. One of the best-represented groups, compared to their described numbers, are the snakeheads.

Snakeheads are jolly attractive animals, which, combined with some serious intelligence (for fish), endears them to me.

It is odd that their family contains only two genera, the Asian *Channa* and African *Parachanna*, who are very similar indeed, both fulfilling the same role of a mid level predator on small fish and aquatic invertebrates within their ecosystems. Now, new species of this group turn up in the trade with comparative regularity (two newly described species, *C. ornatipinnis* and *C. pulchra*, turned up in the aquatics trade almost a month after they were described) as collectors pass the demand for attractive new snakeheads on, and use it to find new fish.

Currently, you can go into some of the better aquatic stores, which cater for people like me who don't care about guppies and get hold of about 4-5 different species of undescribed snakehead, if you are prepared to pay for the privilege! Some species will set you back £150 for a juvenile...

For the rest of us with smaller wallets, but still interested in aquatic life, there are always options. Cypriniformes are a huge group of fish, spreading naturally from North America, to Europe and Africa and all across Asia. Currently, about 2,700

different species of Cyprinid are known, and this number is rising as people explore new areas.

The very cute fish above is a Cyprinid from the family Cyprinidae. That is all we know about it! It's body shape is typical of a small Cryprinid, being fairly thin and spindly in structure. It was originally found by aquarists in India who found it amongst various other species of fish in dealer's tanks. That is the extent of our knowledge about it!

Another unknown fish is this stunning fellow, which was originally known as the Galaxy rasbora, which is the name by which I always knew it. It turned up in the trade about three years ago, and quickly gained a foothold due to its ease of captive care, and superb colours. It is again a Cryprinid, something no-one disputed. But then, some cocky sod on an aquarium forum decided that it is not a rasbora; it was a danio. Now, I am not a fan of name changes: they mess things up massively (I am still trying to come to terms with the change from *Exotic Pets Magazine* into *The Amateur Naturalist*, but hey) and cause me irritation when I refer to the old name, and get some prat saying *"sorry, its not called a Galaxy Ras-*

bora, it now has to be called a Celestial Pearl Danio". Well sorry sunshine, but I don't see you taking the same attitude to the aquarium *Labeo* sharks (like the well known Red Tailed Black Shark), which clearly are not sharks, but everyone refers to them as such! Anyway...

Imports were flooding in from the small locality in which this fish is found, and its natural numbers were beginning to drop. A scientist decided to get out of his chair and describe it, which they did in a slightly slap-dash way, and came up with *Celestichthys margaritatus*, or the Margined Celestial Fish. It then turned out that this fish was fairly easy to breed, and most of the people keeping it had bred them in their own tanks. To stop the natural population from crashing (which it was beginning to do) and then have to suffer the wrath of aquarists (who, for some reason, don't really care about overfishing certain species of large cichlids so they become as rare as hen's teeth, but love a little teeny tiny fish with as much intelligence as paint), imports of the fish were banned. Would this have happened if the fish had not had a scientific name? I think not. "You can only save what you know"...

On a brief note, there is a lot of evidence for this being a *Danio* species, but who cares? A scientific name saved a species.

Let us now look at a group of animals many find disdainful, to put it lightly, the cockroaches. Before we take a look at the different undescribed species though, let us first look at what the hell a cockroach is. They can be grouped into a couple of "looks", each suited for a different habitat and mode of life (NOTE: these groupings are not zoologically significant, merely morphologically so). First we have the large bodied flightless species that spend their time either underground, or under shelter of some sort. We also have the smaller long legged reduced wing species, which lead a more active existence. They tend to run from danger, not fly. The small species with full wings are the most common; indeed, the British species belong to this group. Finally, you have the large winged species. Most of these are partially arboreal, but some become so large that they can no longer fly.

One of many interesting myths about cockroaches is that they will survive through a nuclear war. Well, they will certainly survive better than us, but for an insect, they are nothing special. Most cockroaches have a radiation resistance 6-12 times higher than a human, but animals like flies have even higher resistances. Radiation affects cells only when they are dividing, either by mitosis or meiosis. In humans, cells divide constantly to grow our bodies, repair ourselves or to produce sex cells. In insects, cell division only occurs at ecdysis (skin shedding). Only after they have shed their skin, do the cells begin to divide to quickly grow their bodies before they harden. If a nuclear blast occurred whilst they were hardening, they would be effected. If it occurred whilst the insect was hard, there would be little effect to the insect. However, long-term radiation could hit the insects when they shed, so they are as at risk as us to long-term radiation.

Very few people realise this, but termites, mantids and cockroaches are closely related, and the name Dictyoptera has been erected to name the group. Cockroaches first evolved in the early Carboniferous period, about 359MYA. Mantids evolved from "proto-cockroaches" about 145MYA in the very early cretaceous. So, as *Tyrannosaurus rex* was romping around biting chunks out of Ceratopsians, early mantids were munching their way through small invertebrates on a micro level. Proto-cockroaches (or Blattoptera) are cockroach like insects that the true cockroaches first evolved from. Looking at a large cockroaches' leg, you can see the numerous large spines projecting downwards from the leg. It is not hard to see how these could have evolved into raptorial appendages.

Termites are just social wood eating cockroaches. They probably evolved about 120MYA in the Cretaceous from a similar group to *Cryptocercus*, a small cockroach. Genetic studies have shown that *Cryptocercus* shares more DNA with termites than any other genus of cockroaches. It is the only cockroach to exhibit true social behaviours like caring for its young. I always find it funny when I say "cockroach" to people, and get a repulsed face. Say mantis or termite to them, and they usually hold them with regard and respect in their mind. You now know that they are basically the same thing.

But enough about roaches and their progeny as a whole; we need to look at the species currently in my care. We start with a species known in the trade as the African Bullet Roach. Why? Well, they are from Africa and move as fast as a bullet! They are a small species getting to between 12-17mm as adults and were first collected in about 1997/8 at the base of Mt. Kilimanjaro in Tanzania. They are so easy to keep and breed that they are being used as feeder roaches for larger amphibians and reptiles. Just run that through your head; an unidentified species, being so common and numerous, that it is thrown to toads without a second glance! Madness!

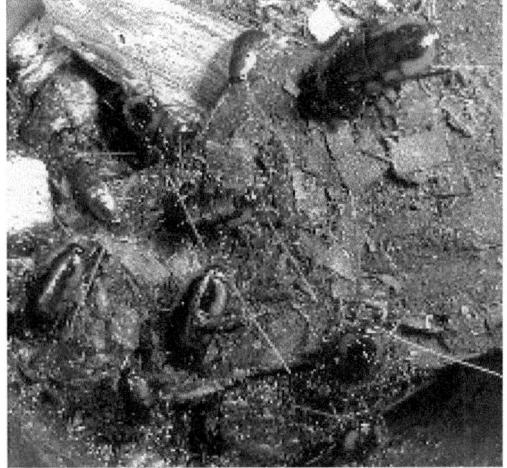

The second species is one that I have yet to photograph properly, due to their small size and fast speed! This is Blaberidae sp. "Kenya". Notice the "dae" on the end of that name and lack of italics. This means that this species is only known to

family level, not even genus (like the above species). Males are an appealing orange colour, but females are a dark brown with reduced wings. They are only small at 10-12mm, but I am really looking forward to breeding them.

Now we look at *Pycnoscelus* sp." Malaysia", a small parthenogenic species (where females can produce fertile offspring without a male) related to the sort-of-pest-sort-of-not *Pycnoscelus surinamensis* which causes a problem with it's size and speed by being able to do well in most conditions (as long as there is a lot of moisture), so it jump's animal tanks easily. They are tiny, which adults rarely reaching 10mm.

A cracking species is *Eublaberus* sp. "Pantanal". This species is related to two common roaches in the hobby, *E. disanti* and *E. posticus*, but has much more black on its head with paler wings. It is found in a more southerly area than the other

two species, and at 50mm, it is longer than the other two species. It is a stunning critter and both nymphs and adults are very showy whilst being rather shy and tending to burrow.

Finally, we have a new colour form of *Eurycotis*

opaca, a large species up to 50mm long. It is related to pest species, but is itself not a pest. It has the build of a running species, but it is much larger than most. Adults have a blaze of colour on the pronotum which looks to me vaguely like a setting sun.

My small collection will soon be growing again when I have the money to buy more roaches/fish/whatever else comes along! All the undescribed species in the CFZ's care are technically breedable, so naturally I will try! The *Channa* sp. "Assam" is unsexed, and I need to find someone with more knowledge on the species than me to sex it, but other than her (I think its a female), all of my other species are breedable. I may soon have more cockroaches than I can care for... never a bad thing!

One of the more important parts of what we are trying to do with the CFZ Outreach Projects is to try and make cryptozoological research accessible to all, rather than just the privileged few.

As Max says, the CFZ is currently keeping colonies of various undescribed species. At present, between us we have colonies of four unknown cockroaches, three fish, and a scorpion, and will be getting more in the coming months. As we breed them, we shall be distributing specimens around to people in the private sector, along with care sheets and record sheets, and shall be asking the new owners to keep records of feeding, growth and behaviour. In this way we hope to be able to gather empirical data on these undescribed species of animal very quickly. We shall then work towards publishing the first scientific descriptions of these fascinating creatures together with an overview of their biology.

If you are interested in getting involved with this project please get in touch by emailing jon@cfz.org.uk or max@cfz.org.uk

Mystery Animals of Flemish Folklore – Part One.
Neil Arnold

On December 5th 2000 a bizarre story made the headlines of the *Volkskrant,* a daily morning Dutch newspaper. It concerned the alleged sighting of a wolf in Belgium, a beast that would become known as the *Waasland Wolf,* a ferocious creature suddenly the prime predatory suspect after a cluster of sheep and chicken slayings in the Flanders region. An extensive hunt was organised to flush out the animal but to no avail, and soon an urban legend was born. Local menus would house 'wolf sandwich', the general public would run rampant through the woods in the hope of a fleeting glimpse of the grey outlaw, and on one occasion, a hunter claimed, by email, to have successfully tracked the beast, killed it and buried it at an undisclosed location. Of course, no wolf was ever uncovered, but if there was only one out there then reporter Herman Koch was quick to jest in the *Volkskrant* of 21st December 2000, (translated as accurately as possible by myself):

'A wolf could never in one day in Walsoorden emerge, two days later in the woods in Heikant, and the following morning again at the Clinge and barely twenty-four hours later at Lokeren.

No, if the wolf already has a map of Dutch Flanders possession, it was undoubtedly a put-up job. Only using a scooter could he pop up in so many places at once – but wolves on scooters heard in fairy tales at home.'

Seven years after the initial fuss, the legend returned with webmaster Karel de Bruyne constructing the website '*Stop de Waaslandwolf*'. He contacted the media after a dog-walker claimed to have seen excrement belonging to the beast, as well as a set of paw-prints along a dyke of the Moervaart. Sceptics were eager to dismiss the reports despite remains of prey being found such as the skull of a sheep. Wolves disappeared from the region after development but rumours have been foggy from the Stekene-Moerbeke-Wachtebeke area.

Wolf is bang van oma's en vindt geitjes ronduit vies

Van onze verslaggever
Bert Wagendorp

AMSTERDAM

Tien miljoen jaar al schuimt de wolf in onveranderde vorm over de aarde, maar wennen doet het nooit. Eén enkel exemplaar laat zich anno 2000 zien in Zeeuws-Vlaanderen, en er breekt paniek uit. 'De wolf van Walsoorden',

was. Van Hilst wordt erg verdrietig van alle vooroordelen. Want als de wolf iets niet is, dan is het wel kwaadaardig of gemeen. Sterk, intelligent, sociaal en een overlever: zeker. Maar tevens doodsbang voor de mens.
Bioloog J. Wensing van Burgers Zoo in Arnhem denkt er precies eender over. Schitterend dier, de wolf, maar ook een 'ban-

dat hij ook wel spreekt van het 'Roodkapje-syndroom' als het om de vooroordelen jegens de wolf gaat.
Terwijl de oma van Roodkapje toch nooit slachtoffer kan zijn geworden van diens eetlust. Wensing: 'De mens staat niet op het menu van de wolf. In de wetenschappelijk bewezen aanvallen van een wolf op een mens, was er steeds sprake van een

In the autumn of '07 protesters strung banners up close to areas where alleged wolf sightings had taken place. On October 7th droppings were allegedly discovered at Etbos by hikers and samples of these were forwarded to the Faculty of Veterinary Medicine. Newspaper *The Sunday* covered the story asking:

'Wolf Waasland back ?'.

Holland is often called the Netherlands, but is in fact a region in the western part of the Netherlands. Sources state that the name Holland, deriving from *holtland* (meaning 'wooded land') first appeared in 866. Those who reside from Holland are known as Hollanders, whilst those from the Netherlands in general are referred to as Dutch. The Netherlands were settled by Paleolithic peoples and strongly influenced by the Celts, Germans, Spaniards, and Romans.

Amsterdam is the capital of the Netherlands, the region has a population of more than 16 million. Despite being heavily populated there are twenty national parks, an abundance of lakes, hundreds of nature reserves and heaths. The Netherlands is

rarely mentioned when it comes to cryptozoology…until now.

In May of 1996 there were rumours that a lion was at large between Warri and De Bilt, in the province of Utrecht. In the autumn of the same year a cougar was said to be wandering around woodlands north of Maastricht, a city in the Dutch province of Limburg, and a leopard was also said to be on the loose in the same year at Zeist, located east of Utrecht. Two years later a puma was reported in forestland at Valkenburg and a bear was sighted at Den Dolder and two years later in Soesterberg. During the August, several walkers claimed they'd seen a lioness at Egmondse duinen.

One of the most famous mystery animals said to prowl the Netherlands, at Veluwe, was the elusive puma, which became known as Winnie. In the June of 2005 police began receiving reports of a large, mystery felid prowling near areas such as Ede, Wekerom and Harskamp. Authorities were so concerned by the reports that large parts of the region were cordoned off and a tracking unit going by the name of Pantera joined the hunt, but to

no avail. Newspaper *De Telegraaf* of June 16th reported:

> *'Ede Municipal Council has hired the Pantera Big Cat Foundation to help catch a puma that is roaming free in the Veluwe forest. The council opted for the Pantera Foundation after its staff claimed they could catch the big cat alive. The foundation has been given from Friday, June 10th , 2005, to Tuesday, June 14th to capture the animal. A mass search by (Netherlands) police over recent days has discovered telltale signs of the cat, and some people have caught fleeting glimpses of it. But attempts to catch it have failed so far.*
>
> *The police had indicated that the puma would be shot dead if it could not be caught quickly.*
>
> *Pantera is based in the town of Nijeberkoop and claims to be the only 'real' big cat reception center in Europe. The foundation was chosen by the town council following consultations with animal welfare groups. The public has been warned to avoid the Veluwe woods near Ede as the search continues.*
>
> *It is unclear how a puma came to be in the Veluwe, a 60-kilometer (36-mile) wooded area that includes the Hoge Veluwe National Park. The police have not received reports of anyone losing their big cat pet, nor have any pumas escaped from a zoo in the Netherlands but police have warned cyclists and ramblers to avoid the woods near Ede because experts say a puma on the loose could be very dangerous.*
>
> *Agriculture Minister Cees Veerman said attempts would be made to capture the animal alive, but, if that failed, it would be shot and killed.*
>
> *The first signs of the puma were detected near Harskamp at the weekend. Over the following days, more traces were found, and members of the public reported sightings. Game wardens also found half-eaten game.'*

The mystery reached some kind of conclusion in the September of '05 when photographer Otto Faulhaber snapped the beast, although in reality the animal was not a puma at all, but in fact a cross-breed of a domestic cat and European wild cat. (See opposite) Why so many witnesses were fooled by a cat nowhere near the size or description of a puma is anyone's guess. The wild cat, rarely sighted in the Netherlands due to its elusive nature, may well have ended all speculation but couldn't explain the handful of reports that described a big black cat also.

In the June 2008 issue (No. 70) of FOAFTALE NEWS: Newsletter Of The International Society For Contemporary Legend Research, Theo Meder of the Meertens Institute wrote:

'THE SEARCH FOR WINNIE THE PUMA: WILD ANIMALS IN A CIVILIZED ENVIRONMENT

In the summer of 2005, a wild animal

Dutch police hunting Winnie the Puma

was spotted on the Veluwe, by a tourist in moorland and forest area in the east of the Netherlands. In the national media, the animal was soon presented as being a puma - a species not indigenous to Europe. With land rovers and even a helicopter, the police tried to hunt down the predator, determined to shoot and kill. Amongst others, left wing politicians and the Society for the Prevention of Cruelty to Animals protested: the animal should be caught, not killed. Mr. Arno van der Valk from the Panthera Foundation claimed to be able to stun and catch the puma alive, and the authorities agreed to leave the job to him.

Meanwhile, so-called puma tourism came into being, and Winnie the Puma was spotted by people all over the Veluwe, sometimes on several places at the same time. The puma became such a hype in the media that the stories started to create a reality of its own. At first, no one bothered to listen to the few experts (biologists and folklorists) who suspected the whole story was a contemporary legend. After a while, when Van der Valk proved to be able to shoot photos and video footage of a supposed puma, but failed to stun

the wild animal itself, more and more people turned sceptical. Many jokes commenting on the hoax started circulating.

It all turned out to be a case of WY-BIWYS (What You Believe Is What You See) and quasi-ostension (a misinterpretation of facts in reality, based on existing narrative scenarios). Folklorists recognize the international contemporary legend called Big Cats (Panthers etc.) Running Wild. It is just another version of the tale of the dangerous and uncontrollable monster walking through our civilized and peaceful backyard. How often do these tales emerge, and how modern are they, as far as the Netherlands are concerned? What parties are involved these days? And why do people keep falling for the tales time and again?

The story of Winnie the Puma ended with photos of a large tabby cat (2005) and got canonized by a chapter in Peter Burger's new collection of contemporary legends (2006) and a huge steel piece of art on the Veluwe by Maarten de Reus, called Cage-With-No-Puma-In-It (2007).'

Biologist Marc Damen of Arnhem Zoo supported the wild cat theory. He said several dozen cats had gone wild live in the Veluwe.

As for the idea of a puma in the national park, Damen said: *"I still have not seen any evidence of this."*

In 2006 website *WaarMaarRaar* ran a brief report on the possible sighting of a kangaroo in the vicinity of Hooge Zwaluwe. Despite a handful of sightings and a police search no hopping intruder was found.

And what of the giant fish, serpents and out of place whales said to haunt the depths of the Dutch waterways? Well, they are for another time…

ANOTHER MYSTERY CAT IN HONG KONG

by Jonathan Downes and Richard Muirhead

As regular readers of this magazine over the past 50 years will know, the current authors have an enduring interest in the fauna – both crypto and otherwise – of the erstwhile British Crown Colony of Hong Kong, which borders upon the obsessive. Both of us spent our childhoods there, and it was whilst surrounded by the incredibly rich fauna of the territory that we both acquired our lifelong fascination with the machinations and diversity of the natural world.

In a desultory fashion we have been working on a book about the mystery animals of Hong Kong since 1993, and will eventually get round to finishing it. Our files containing source material on the subject seem to threaten to defy the laws of physics and take over the CFZ archives on occasion. When one of us finds something new, we telephone the other in a highly excited manner, and occasionally – like now – we feel that we have to share our findings with the Animals & Men readership.

The thing, which has excited our imagination so much on this occasion, is a series of peculiar events during the mid-summer of 1910.

It is not in doubt that 3 species of wild felid have been reported from the region in recent historic times. They are:

- *Pironailurus bengalensis* – the Bengal leopard cat, which is the only species extant in the Special Administrative Regio
- *Panthera pardus* – the leopard. This was always the rarest of the local felids. Writing in 1952, G.A.C. Herklots reported:

"*The visit of a leopard to the Colony is a much rarer event than that of a tiger. On 20th December, 1931, one was shot near the village of Chung Pui to the north of the Pat Sin range.*"

The photograph of this skin can be found in the *Hong Kong Naturalist*. Another specimen was shot over 20 years later in the late 1950s.

The Hong Kong Naturalist.   Vol. I No. 1.

Photograph of skin of Leopard, *Felis pardus*.

Plate II.
Printed by S.C.O. Post.

- *Panthera tigris* – the tiger. The history of tiger visits to Hong Kong was covered in some depth in issue 9 of this magazine back in the Spring of 1996. Tiger reports continued throughout the 50s, 60s and

1970s, although the last certain record took place in 1947 when no less a personage than the Bishop of Hong Kong reported seeing a specimen walking across his garden.

However, new evidence has emerged which suggests that a 4th cat species may have occurred in Hong Kong on at least one occasion.

For the moment we shall discount reports of what appear to be a black panther which was seen in the environs of Stanley Internment Camp during the early days of the Japanese occupation. As written elsewhere, the tiger which was shot by Japanese soldiers, or – to be more accurate – by either Formosan or renegade Indian troops in the employ of the occupying imperial Japanese forces.

As described in Jon Downes' book *Monster Hunter* (CFZ Press, Exeter 2004) there is enough evidence to suggest that this tiger was not a specimen of *F. t. amoyensis*, but an ex-circus creature brought into the colony and liberated by the Japanese psychological warfare experts as an attempt to destabilise whatever local support there was for the King-Emperor George in favour of the God-Emperor Hirohito, by utilising an old Syno-Japanese (in fact Pan oriental) piece of folk belief which implies that the arrival of a tiger in a neighbourhood where a tiger has not been seen before is indicative of a change of emperor. The fact that the "black panther", and – if memory serves us well – a wolf – was seen at the same time as the unfortunate tiger whose pelage can still be seen at the Tin Wa Temple in Stanley village, would suggest that they had the same origin, and cannot really be considered valid supplicants to the Hong Kong mammalian species list.

This new specimen from 1910, however, is something completely different – in all senses of the word.

The *South China Morning Post* of Monday June 27th 1910 reads:

> "Residents in Kowloon ought in future to rest peacefully at night, for the prowling "tiger" has at last been killed. This was the news which reached our office last night.

It appears that on Saturday afternoon several parties set out into the New Territory with the avowed intention of bringing home the skin of the "tiger" if there was such an animal anywhere in the reach of a gun.

One of these parties consisting of 6 men, set out late on Saturday afternoon. From information they obtained from outlining villages, they learned that the beast was in the habit of feeding in the neighbourhood of Tai Om between 4 and 5 in the morning. The hunters accordingly concealed themselves near the place and waited with commendable patience. They were rewarded for their long wait, for at half past four on Sunday morning it made its appearance. Two shots were fired at it, one striking it in the neck and the other in the back. Both shots took effect, one of the bullets entering the spine, and the beast at once collapsed and in a very short time was dead.

In its death struggles the animal dug a hole in the ground nearly 3 feet deep. It was a powerful beast and measured 5feet 1inch in length without the tail. It stood about 3 feet high. Its skin was of a dark brown hue, and it did not resemble a tiger. It is thought to belong to the panther specie. [sic]

The two successful shots were fired by Sergt. Devny and Mr. Gast. They are to be congratulated on the success of their hunting expedition, and they will have the satisfaction of having relieved the minds of the residents in the outlying villages of the New Territory. They left the animal in the New Territory, we

understand, for the purpose of having the skin preserved. What will be done with it afterwards has not
been decided. It will make an interesting addition to
the local museum".

On the same day the *China Mail* printed a chattier account of the
expedition which includes an interview with Mr. Gast who – in
turns out at the time - was a warder at Victoria Jail:

> *"'What colour was its skin?' asked our representative.
> 'Brown,' replied Mr. Gast. 'I am sure it was not a tiger'
> he added, 'Its more like a kind of jaguar or panther.
> There were no stripes at all on his body.'"*

Two days later the same paper celebrated the death of the mystery
creature in bad poetry:

> "At last he's caught, that awful beast!
> No more on Chinese pork he'll feast.
> At Tai Om asleep he lies,
> A glassy brightness in his eyes"

On 28th June, the *South China Morning Post* published another
account of the hunt concluding with the lines:

> "The old native was unable to say what the animal
> really was, but expressed the opinion that it was a
> yung-kan, or in other words, a wolf. We understand,
> however, that the animal was twice the size of an or-
> dinary wolf and weighed no less than 186 pounds.
> The natives of the village took the carcass after it had
> been skinned, although it is likely that had the hunt-
> ers so desired they could have disposed of it at ₱2 a
> pound. The skin is being cured but none of the party
> at present know who will eventually become the lucky
> owner."

This is just another confusion because although the weight given
is large for a wolf, it is just about possible. There is, however, a
problem: there are no known wolves in Hong Kong, and as far as
is known the only wild canids in the region were the Chinese race
of the red fox *(V. vulpes hoole)* and the red dog *(Cuon alpinus)*
both of which are considerably smaller than wolves, and both of
which are - coincidentally - now extirpated from the region.

Another local newspaper, the *Hong Kong Daily Press* treated the
whole affair facetiously on the 29th June, with the sort of heavy
handed humour that our grandparents found funny, and which
these days is just seen as embarrassing if its done.

The big question is what the hell was it?

THE OWLOON
"TIGER."

Bagged at Last.

Shot at Tai Om.

Residents in Kowloon ought in
future to rest peacefully at night
for the prowling "tiger" has at
last been killed. This was the news
which reached our office last night.

It appears that on Saturday after
noon several parties set out for the
New Territory with the avowed intention of bringing home the skin
of the "tiger" if there was such an
animal anywhere within reach of a
gun.

One of these parties, consisting
of six men, set out late on Saturday afternoon. From information
they obtained from residents in the
outlying villages they learned that
the beast was in the habit of feeding in the neighbourhood of Tai
Om between four and five in the
morning. The hunters accordingly
concealed themselves near the place
and waited with commendable
patience.

They were rewarded for their long
wait, for at half-past four on Sunday
morning it made its appearance.
Two shots were fired at it, one
striking it in the neck and the other
in the back. Both shots took effect.
one of the bullets entering the spine,
and the beast at once collapsed and
in a very short time was dead.

In its death struggles the animal
dug a hole in the ground nearly three
feet deep. It was a powerful beast
and measured 5 feet 1 inch in length
without the tail. It stood about
3 feet high. Its skin was of a dark
brown hue, and it did not resemble
a tiger. It is thought to belong to
the panther specie.

The two successful shots were fired
by Sergt Devney and Mr. Gast.
They are to be congratulated on the
success of their hunting expedition.
and they will have the satisfaction
of having relieved the minds of the
residents in the outlying villages of
the New Territory.

They left the animal in the New
Territory, we understand, for the
purpose of having the skin preserved. What will be done with it afterwards has not yet been decided. It
would make an interesting addition
to the collection at the local museum.

RUBBER NEWS

There is, however, one species that – with one important exception – fits the bill almost exactly. It is the Asian golden cat (*Pardofelis temminckii*). It is found throughout great swathes of South-east Asia from Tibetan Nepal through Southern China and India to Sumatra.

Indeed during one of the CFZ expeditions to Sumatra putative orang pendek hair samples analysed by our buddy Dr. Lars Thomas in Copenhagen turned out to be from this species. However, not only are there no records of the golden cat from Hong Kong, but the body length (sans tail) of this species is three foot, not over five.

I have also not been able to ascertain any wild records from Guangdong Province, however recent WWF reports do mention it as turning up in food markets.

However, here it should be noted that none other than Bernard Heuvelmans suggested that there could be a gigantic version of a related species – the African golden cat – in east Africa, and that this was most likely to be the specific identity of the fabled mngwa or nunda; a mystery cat whose predatory instincts are legendary.

None of the present authors are prepared to go as far as saying that this is what had occurred in Hong Kong during the summer of 1910, but it is a scenario which is worth bearing under consideration.

Before one leaves the subject of the mystery cat of 1910, it should, perhaps, be noted that there are at least 2 Fortean aspects to the case. Firstly, of course, is the fact that mysterious puma coloured cats that appear in places where they really have no right to be are so common within the canon of Fortean zoological law really need no introduction. Was this merely an early example of such a now-commonplace Cryptozoological archetype?

The second Fortean zoological aspect to the case is more subtle. It was John Keel, we believe, who first introduced the Fortean public to the concept of window areas; places where there are more strange occurrences than one would expect. During our research we have described a number of such things and places, most notably the events in south Devon during the summer of 1997 (and intermittently before) as described by Jon and Nigel Wright in their book *The Rising of the Moon* (now available in a revised edition published by Xyphos Books, Bangor, NI, 2007), and Jon's book *The Owlman and Others* (CFZ, Exeter 1997) which describes the window area of Falmouth Bay during the summer of 1976.

Something similar seems to have happened in Hong Kong during the summer of 1910. For example, in June and July alone, the following creatures were reported:

- ·A giant tadpole like creature with a square head and a rat like tail found at Pai tan (*South China Morning Post* June 18th)

- A sea serpent with 'a very large head with an enormous mouth, ornamented with large teeth of about 1 –2 inches in length, its body was about 19 foot long and of a silvery colour underneath and a dirty grey colour above' (*South China Morning Post* July 21st)

- A 'species of python with innumerable legs' and 'a peculiar species of rat observed after periods of drought' (*Hong Kong Daily Press* June 29th)

There was even a story of a giant crocodile. The *South China Morning Post* on July 14th reported :

"Following the tiger story came the tale of a crocodile in Kowloon Bay. A well stretched imagination gave out that the creature was alive and 40 feet long, and disappeared from a view with a long boathook in its ribs. We are now reliably informed that a small crocodile, about 4 or 5 feet long, has been washed ashore at Mataukok. It had a cord attached to one of its hind legs indicating that it had been in captivity. The animal had probably died aboard some ship and had been thrown overboard."

As always seems to be the case in Fortean investigations, the whole affair throws up far more questions than it does answers. The mysterious water creature sounds very much like a triops. The two stories from the Hong Kong Daily Press sound like examples of the aforementioned heavy handed Edwardian humour, but both the 19 foot sea serpent and the mystery cat which has taken up the vast bulk of this article remain very real, and potentially tangible mystery animals.

We are carrying out various avenues of investigation including trying to trace the descendants of Mr. Gast in the vain, but not impossible, hope that in someone's attic somewhere is a very interesting pelt.

CORRESPONDENCE.

THE SEA SERPENT.

To the Editor of the "Morning Post."

Sir, - I have refrained from writing to you before owing partly to diffidence and partly to the incredulity with which certain stories told recently by newspaper occasional correspondents have been received. Fortunately, I have at least a dozen credible witnesses who can vouch for the truth of my story.

On Sunday last, at about 5 p.m., in a small bay opposite the Junk Bay Flour Mills, we had a most exciting experience with what was apparently a sea-serpent. The reptile, whatever its species, had a very large head with an enormous mouth, ornamented with large teeth of about 1 to 2 inches in length, its body was about 16 feet long, and of a silvery colour underneath and a dirty grey above.

Some of our party were going ashore in a dinghy when they caught sight of the reptile in the water, and the gentleman who was rowing promptly struck at it with his oar, and apparently stunned it. The ladies in the dinghy naturally shrieked and refused to go on shore, so the dinghy was put back to the launch. The sea-serpent had apparently recovered from the blow it had received, for it seemed to turn round and follow the dinghy to the launch, which it perhaps had mistaken for its aggressor. On reaching the launch it was found, however, that the stern painter of the dinghy had entangled the poor snake and dragged it along in its wake.

Perhaps some of your readers saw the same thing on Sunday last, as there were five launches in the vicinity.--Yours, etc.

GEORGE WASHINGTON II.

P.S.—I ought to have mentioned about that the poor reptile was dead, and had apparently been washed down from the hills during the recent rains.

SOUTH CHINA MORNING POST JULY 21ST 1910

THE MOA YOU KNOW

Oll Lewis

Moa were truly fascinating creatures. These flightless birds, native to new Zealand, could grow to around 3.7 meters (12ft) tall with their heads held aloft in the manner of an emu or ostrich as they are depicted in most reconstructions of their skeletons (although in practice moa rarely would have raised their heads in this manner and would have kept them level with their bodies in the same was a kiwi does).

These large and robust leaf-eating birds had only one predator, the Haast's eagle, and were common on both North and South Island before the arrival of man around 1300 AD. The Māori regularly hunted moa for food and other uses and also cleared large tracts of forest and this led to the supposed extinction of moa and the Haast's eagle by around 1400 AD.

As any cryptozoologist knows though, just because an animal has been declared extinct or is only known though fossils or its remains it doesn't necessarily mean that an animal *is* extinct. For an animal to be declared extinct it merely means that there is no scientifically verified evidence of the species survival for 50 years. This can just mean that the species is low in numbers and good at hiding or that scientists were just looking in the wrong place.

Another New Zealand native bird, the Takahē,

On the 20th of January 1993 former SAS instructor, Paddy Freaney, and 2 of his friends, Sam Waby and Rochelle Rafferly, were tramping in the Craigieburn mountain range on South Island when they saw a 6 foot tall bird. Freaney immediately recognised the bird as a moa. He described the animal as being 1 m tall and having a long thin neck, also around 1 m that tapered to a small head. The animal was covered with red-brown feathers apart from the legs and feet below the knee.

The trampers startled the animal and it ran across a stream, but Paddy gave chase and was able to take a blurry photograph of it when it was around 40 m away. Paddy also took photographs of what he believed to be tracks although these have never, to my knowledge, been publicly released.

When the story broke of the moa sighting the New Zealand government's department of conservation expressed an interest in funding an expedition to the area to search for evidence of possible extant moas and the photographs and negatives were analysed by the electrical and electronic engeneering department of the University of Cantabury.

The University confirmed that the photograph was of a large bird, however a former phd student at the university, palaeoecologist Richard Holdaway, went public in saying he thought the picture might be of a red deer. Because of Holdaway's assertion the department of conservation pulled out of any further investigation of the moa sighting.

Far be it from me to cast aspersions upon Mr Holdaway's opinion, but if the trampers witness

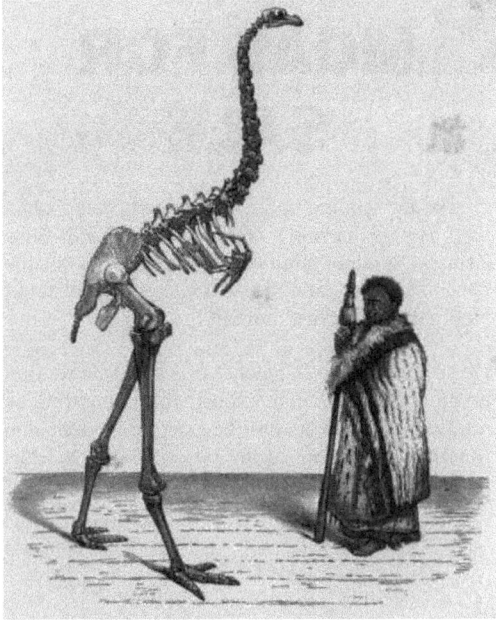

was thought to be extinct after not having been seen for 50 years, but was rediscovered by Geoffrey Orbell in the 1940s. Orbell was a keen tramper, or bush walker, and could be described as a cryptozoologist.

Orbell was convinced that, despite having been declared extinct, the Takahē was still extant in New Zealand and kept a close watch out for anything unusual when he was bush walking and eventually he came across a set of animal tracks he had never seen before in a remote valley near lake Te Anau. He followed these tracks with three other people and found a Takahē at the end of them. If only all psudo-cryptids were as obliging as the Takahē!

The fact that the Takahē was rediscovered certainly can not be used as evidence that at least one species of moa may still be extant in New Zealand, but it does illustrate just how easy it can be for a species with a low population size to evade detection in New Zealand's wild places. An encounter with a moa is alleged to have occurred in 1993 in just such a wild place:

PADDY FREANEY   SAM WABY   ROCHELLE RAFFERTY

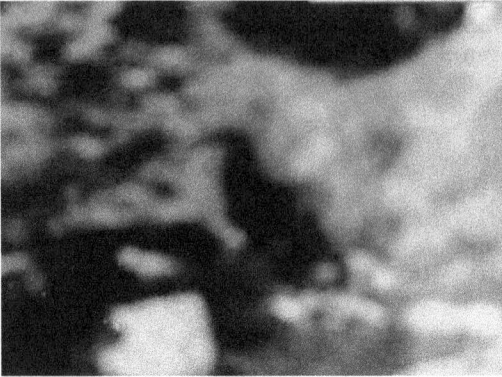

statements are not false then I would have thought that a bipedal deer covered with feathers as tall as a man might have been unusual enough to merit further study.

As a result of Holdaway's, perhaps somewhat creative, explanation of the photo the tide turned on Paddy and he found himself being accused of making the whole thing up, this even went as far as claims by news papers that he was part of a liars club but the papers in question soon retracted this statement.

Unfortunately, the photo is too blurred to say for certain what it is a photograph of but the subject does have a bird's shape and according to the people who analysed the photo it was not a card-board cut out.

If it is not a photograph of a moa it could be a photograph of another species of bird that has been misidentified rather than a deliberate hoax, but Paddy Freaney has always remained resolute that he is telling the truth about his encounter and has conducted several small scale expeditions in the area in the hope of finding a moa and proving his story. Sadly to date they have been unsuccess-ful.

Did the three trappers see a moa alive in the 20th century? Might at least one species of moa have survived into modern times? Perhaps it is just within the realms of possibility, but at present there is no conclusive proof.

MOAS FOR SALE

One of the leading lights of the newly established CFZ Bloggo network is Tony Lucas from New Zealand; already familiar from his articles for the CFZ Yearbook and for *Animals & Men* about New Zealand Cryptozoology.

He seems to have somewhat of a knack in finding cryptozoological, more specifically moa-related artefacts on sale in on-line auction houses. On March 22 he wrote to me twice. The first letter read:

> "I've had this guy get in touch with me and is selling a Moa feather
>
> The guy rang and spoke with me to-night, the feather, that was in a first edition Maori bible that had been locked in a metal trunk since 1900, has apparently been authenticated by Te Papa Museum in Wellington as the genuine article. I was wondering if the CFZ museum would be interested in purchasing this item"

Well, despite the fact that we are skinter than we have been in a long time, with enormous printer, garage and council tax bills on our plate, I was indeed interested, yes.

But then came a second e-mail:

> "Just done some research and Don't think this is a Moa feather, if you look at the bottom of the second pic of the feather you can see the split where a second spine came off, Yes there were two feathers attached, The only bird to have such feathers after extensive re-search is - You guessed it an EMU. Sorry to get your hopes up Its not the genuine article by the look of it so forget

it. Feel a bit of a dumb ass now. Still who knows one day perhaps the real thing."

Eventually, nearly two months later the affair reached its dénouement:

As you may well remember, if you have been keeping up with the blog, a few weeks ago I wrote to Jon regarding a Moa feather that was for sale on the Internet auction site "Trademe".

I decided to conduct an investigation into whether this was a genuine Moa feather, which if it was would have proven the Moa was still around in the late 1800s, or whether it was a very similar Emu feather. After some investigation I found an anomaly at the bottom shaft of the feather which

HEINRICH HARDER

seemed to indicate a secondary feather which would definitely indicate emu as it is the only bird in the world which has this double feather configuration. My suspicions aroused I tell the seller to take it to the museum which they did and visually identified it as Moa, they also wanted to send a small portion away to be DNA tested. I advise the seller strongly to do so which he duly did.

I decided to follow this up the other day and wrote to the seller, with the merest slimmest of hope that it was indeed a Moa feather.

I received his reply last night, and re-grettably for both him, he was trying to get the money together to buy a house, and I it turned out to be a 100-year-old emu feather.

The one thing I've learned from this is not to take things at face value but to investigate everything thoroughly, there was apparently an American buyer inter-

How does the bloody man do it?

It is becoming ever more obvious that various members of the CFZ bloggo team have what Charlie Fort would have described as "wild talents". Richard Muirhead and Nigel Wright (for example) no sooner have to walk into a public library before they unearth a hitherto unsuspected treasure trove of data.

Max Blake is equally lucky with tropical fish shops, and it seems that our resident New Zealander Tony Lucas has a similar Lovejoyesque power over auction houses.

Cop   a   load   of   this.

He wrote to me saying that this was something in which we might be interested, and the answer is that of course we are (even

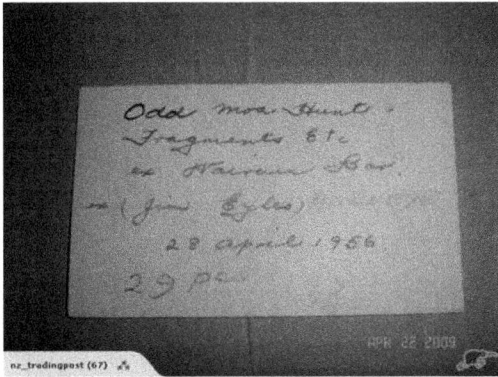

1956 By Jim Eyles who was one of New Zealands best Archeologists at the time.

There are 29 pieces all up. There is a card inside the box which reads Odd Moa Hunt Fragments Etc Ex Wairau Bar (Jim Eyles) Hawkeye 28 April 1956 29 pcs. Jim Eyles was a well known Moa Hunter. Later in life he was the Director of The Nelson Provincial Museum. Please feel free to ask me any questions you have regarding this item and i will answer them as best as I can.

To describe Jim Eyles as a "well known moa hunter" is an understatement. To quote wikipedia: James Roy (Jim) Eyles OBE (1926–2004) was a New Zealand archaeologist.

Born in 1926, Jim spent his early years living with his family at the Wairau Bar near Blenheim where, in 1939 as a schoolboy, he discovered early human skeletons and associated artefacts including necklaces, stone tools and moa egg.

Jim discovered several more burials on the bar, and assisted with the excavations carried out by Roger Duff from the Canterbury Museum. The site is regarded as one of the oldest and most important archaeological sites in New Zealand.

So this item is actually quite an important piece of cryptozoological history. We have no money, but of course we are bidding on it. If anyone fancies donating a few quid towards the purchase price we would be very grateful..

though we still have some financial problems, even though the worst have now been alleviated).

The description is as follows:

This auction is for some RARE Moa Fragments and other items found Ex Wairau Boulder 28 April

A WHALE OF A TALE

L ast summer when the CFZ expedition-ary force were in the mountains of southern Russia, Richard Freeman and the gang used to while away the long nights by talking about cryptozoological matters with their guide and colleague Grigory Panchenko. One of the stories that Grigory told them was of a whale which had turned up dead on the beach on the island of Sakhalin. Sakhalin is a

strange island with a strange history. I first heard about it in the late 1960s when I read a book called *Biggles Buries a Hatchet*. No, this was not some strange prescient piece of prediction about the day that Corinna's collie pup buried some of Graham's DIY tools in the rose bed. No, it was the story of how Major James Bigglesworth, all-round good egg and hero of the British Empire, and his gallant crew fly to Sakhalin in order to

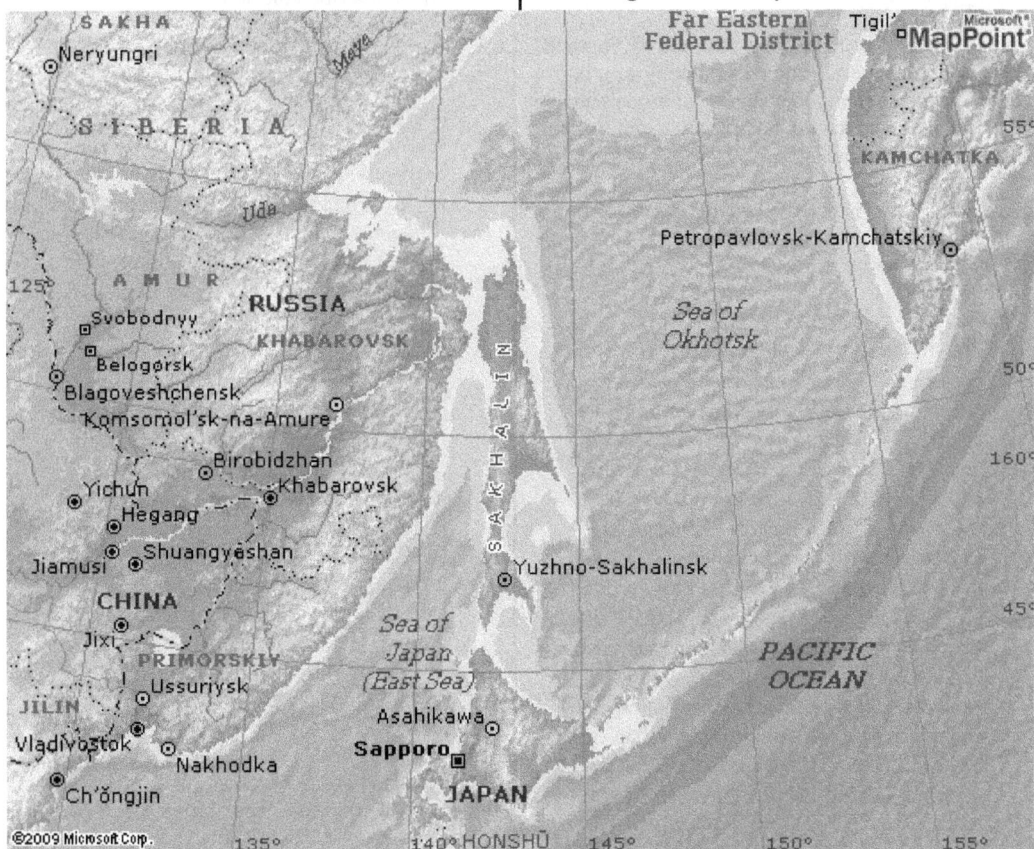

rescue their erstwhile nemesis Erich Von Stalhein from a Soviet prison after Von Stalhein unaccountably fell foul of the Soviet leadership. This got me quite interested in the island of Sakhalin.

The island has been the subject of territorial disputes between the Russians and the Japanese for centuries and both geographically and culturally it is closer to Japan than Russia. Japan unilaterally claimed sovereignty over the whole island in 1845 and ten years later both countries signed the Treaty of Shamoda which declared that both nationals could inhabit the island: Russians in the north, and Japanese in the south, without a clear boundary between. The island didn't become completely Russian until the end of August 1945, at the end of the Second World War when the Soviet Union finally took over control of the island, and Japan finally renounced claims of sovereignty in the Treaty of San Francisco (1951). However, the cultural and geographic schizophrenia of the place has meant that my ears are pricked up whenever I hear it mentioned.

When Richard and the gang returned to England, Grigory promised to send them a copy of a paper that he had written about this mysterious carcass that he believed could have been from a very primitive whale species.

Eventually, the paper arrived on the CFZ computers in the early spring of 2009.

It tells how:

> "In the second half of September 86, Sergei Litvinyuk, while on a geological expedition in the south western extremity of Sakhalin island, observed and photographed a large sea animal carcass washed ashore near the town of Gornozavodsk on the coast of the Strait of Tartary. After consulting with local inhabitants, Litvinyuk decided that the carcass most probably belonged to a species of big whale (Ziphiidae). The carcass lay in a surf zone making it difficult to obtain fragments, while the expedition jeep was not powerful enough to drag the several ton body up the beach. These circumstances, plus Litvinyuk's belief that the specimen belonged to

a known species, explain his failure to obtain the mandible or at least some teeth from the carcass".

The paper which was written by Grigory and Sergei went on to describe the size of the creature and concluded that it was about 8 metres in length. The "integument was almost totally destroyed by the sea" and the layer of what looks like skin the photograph is in fact a light coloured layer of tough subcutaneous fat.

The few remaining fragments of skin looked smooth and dark and prompted the two authors writing for the Russian Society of Cryptozoology to conclude that the specimen in life had been uniformally black in colour. They went on to describe how the flukes had not been preserved, and some vertebrae could be seen along the last part of the tail, covered only by shreds of muscle tissue.

They then described the skull:

> "The long beak, with over 1 metre long jaws, is merging gently with the rest of the head. The beak is somewhat flattened in vertical direction, but not so much on the sides as is the case in most members of the Ziphiidae. The most remarkable feature, regrettably not well seen in the photo, was numerous teeth on both jaws. According to Litvinyuk, the teeth, measuring about 10 centimetres in length, straight, conic-shaped, round in cross-section, forming one row, were not set closely together, expect rather worn out front teeth (probably one or two pairs). The latter, about 6 centimetres in length, sat on the very tip of both jaws, set closely together and resembling incisors of hoofed animals".

They went on to describe that there were no direct signs of putrefaction, probably explained by the extremely low temperature of the water at the time.

The two authors then went on to examine a number of primitive and little-known whale species and then produced a series of arguments, most notably a prominent "crop" which "may be explained by a peculiar angle between the ramus and body of the jaw".

They go on to say:

> "Among the cetaceans, such a "crop" is only known in some species of the extinct suborder Archaeoceti."

And then they write:

> "Dentition (except the extreme cheek teeth that retained saw-like edges) was also similar: straight and cone shaped, set in one row and not very close together. If in the carcass the upper jaw really had teeth on the front edge, it means they were on Os praemaxilla, which in modern whales is devoid of dentition, and is characteristic of Archaeoceti and to a lesser degree of the extinct family Squalodintidae, the most ancient and primitive of Odontoceti."

They conclude that if all of their assumptions are correct, the specimen cannot belong to any known family of extant cetaceans, but admit that if they were only partly correct, the carcass was that of a beaked whale.

And a beaked whale it was. Richard and I identified it as such, and when we passed the evidence on to Darren Naish he, sadly, agreed with us.

He wrote:

> So what was the Sakhalin Island carcass? For starters, it was, without doubt, that of a cetacean. Panchenko and Litvinyuk argued that such groups as kogiids and delphinids could be excluded from comparison, and I agree. But what about Litvinyuk's original identification of the carcass as that of a ziphiid, or beaked whale? Based on the apparent presence of a number of teeth in the upper jaws, and of an upper jaw that overhung the lower, the authors concluded that a ziphiid identity could be excluded. They also thought that the presence of a 'crop' in the neck excluded a ziphiid identity.

> By 'crop' they were referring to a large convexity in the throat: their use of the term 'crop' is rather confusing given that (among tetrapods) crops are unique to birds, and there is no structure in a cetacean, or indeed in any mammal, that can be referred to by this term. Finally, because the large incisor-like teeth were at the tips of the upper jaws, this cetacean must (so the authors argued) have had a premaxillary dentition, and they also noted that this was inconsistent with a ziphiid identification.

And Darren concludes:

> As is obvious from the photos that were taken at the time, the carcass *is* that of a ziphiid. Furthermore, the distinctive shape of the long, superficially duck-billed jaws shows that it's clearly a fourtooth whale (*Berardius*), almost certainly Baird's beaked whale *B. bairdii*.

So, sadly, another putative late-surviving archaeocete bites the dust.

CFZ NEWS

ALL CHANGE

Things *are* changing at the CFZ, and we believe that they are changing for the better. Although we have had a web presence since 1997, and a proper website since 1998, it is only now that we are becoming a truly web-based organisation.

However, `web-based` doesn't mean what it does in most cases. Usually when one reads that an organisation is becoming `web-based` it means that it is well on its way to becoming a website and nothing much else. Well that just is not going to happen with the CFZ.

However, it is about time that we started to utilise the power of the Internet, and so - with this issue of *Animals & Men* - the beginnings of the change to what could vulgarly be called CFZ v.2.0 can be seen.

Except it is nothing of the sort. If anything it is CFZ v.5.0.

CFZ v1 (1992-1996) was run pretty much single handedly by my ex-wife Alison and me. It put out *Animals & Men* and maintained an archive, and generally acted as a clearing house for information.

CFZv2 (1996-2002) was much the same, only a bit more laddish, and was run by Graham, Richard and me.

CFZv3 (2002-2005) was where it began to become a proper field study organisation with high profile and productive research both here and abroad.

CFZv4 (2006-2009) was where the field studies of v.3 were augmented by an upsurge in publishing and a move towards a family-friendly image.

And now, in our seventeenth year, the year we publish our fiftieth book, the fifteenth year of this magazine, and the year I attain my half-century it is changing again.

The idea of calling what we do CFZ v5.0 is just vulgar, so it will not be mentioned again, but it is a useful way of illustrating how far we have come from our very humble beginnings.

However, the CFZ bloggo has turned into a daily magazine bringing interested parties up to date with the latest news in the fields of cryptozoology and the other disciplines (arcane and prosaic) with which we deal.

This has allowed *Animals & Men* to grow and turn into the periodical that you are reading now. Our sister magazine *Exotic Pets* has ceased to be and has been reborn into another perfect bound periodical like this one, but now rechristened *The Amateur Naturalist*.

Both magazines will continue to come out three or four times a year, but as we feel that the content is more important, rather than keeping to spurious deadlines in order to please advertisers who usually didn't pay their bills anyway, we stress that all subscriptions are for four issues rather than for twelve months, and so not to worry if - like on this occasion - there are long gaps between issues.

The CFZ yearbook will continue to be published annually, and we are also in the early stages of planning an academic peer-reviewed cryptozoological journal as well,.

The challenge of coming up with nearly 4,000 bloggo postings (not counting news items) a year, plus filling 1000 pages of magazines and yearbook are considerable, but we feel that we are going to be able to rise to the challenge, and continue making the CFZ an organisation like no other that has ever existed in the field of cryptozoology.

WEIRD WEEKEND 2009

This timetable is ridiculously provisional, and that the CFZ take no responsibility for disappointment caused by the non-appearance of any of the advertised speakers

THURSDAY

7.00 p.m Cocktail party at the CFZ
Myrtle Cottage, Woolsery, Bideford, North Devon EX39 5QR

FRIDAY

Noon - 5.00 p.m Open Day at the CFZ
Myrtle Cottage, Woolsery, Bideford, North Devon EX39 5QR

Community Centre, Woolsery, Bideford, North Devon
Doors open at 6.00

7.00 – 7.15 Introduction
7.15 – 7.45 OLL LEWIS:
The Kraken and the Colossal Misunderstanding
7.45 - 8.15 JULIAN VAYNE:
A cabinet of curiosities from North Devon Museums
8.15 - 8.45 BREAK
8.45 - 9.30 JON McGOWAN:
Big Cats in Britain - exposed!
9.30 - 10.00 BREAK
10.00 - 11.00 TIM MATTHEWS:
Crop Circle Confusion

SATURDAY

Community Centre
doors open at 10.00

11.30 – 11.45 JON DOWNES + RICHARD FREEMAN:
An introduction to cryptozoology
(ALL AFTERNOON JOANNE CURTIS:
monster making for kids)
11.45 – 12.15 MAX BLAKE:
Unknown animals in Pet Shops

12.15 – 1.15 PAUL VELLA:
The Minnesota Iceman
1.15 - 1.45 BREAK
1.45 – 2.45 ALAN MURDIE:
Forteana from Colombia
(PAUL VELLA: Bigfoot for kids)
2.45 - 3.15 BREAK
(KIDS: Mad Hatter's Tea Party)
3.15 – 3.30 QUIZ
3.30 - 4.30 ANDY ROBERTS:
The big grey man of Ben McDhui
4.30 - 5.00 Break
5.00 – 6.00 JAN BONDESON: *The basilisk*
6.00 - 7.00 DARREN NAISH:
British big cats in deep time
7.00 - 7.30 Break
7.30 – 8.30: NEIL ARNOLD:
Zooform Phenomena - monsters amongst us
8.30 – 8.45 CFZ AWARDS
8.45-9.15 Break
9.15 – 10.15 TBC
10.15 - 11.00 TIM THE YOWIE MAN:
Yowies - Australia's Bigfoot

SUNDAY
COMMUNITY CENTRE
doors open 10.30

12.00 – 1.00 MICHAEL WOODLEY:
A proposed system of taxonomy for cryptozoology
1.0 – 1.30 GLEN VAUDREY:
Mystery Animals of the Western Isles
1.30 - 2.00 LIVE: Sitting Now (Discussion and Panel) Presented by KEN EAKINS
(KIDS: Monsters are real – Jon Downes/Richard Freeman 30 mins)
(ALL AFTERNOON JOANNE CURTIS:
monster making for kids)
2.00 – 2.30 BREAK (KIDS: Treasure Hunt)
2.30 – 3.30 NICK REDFERN: *Stalin's ape men*
(OLL LEWIS: Lake Monsters for kids)
3.30 - 4.00 BREAK
4.00 – 4.45 RONAN COGHLAN:
Atlantis and other lost continents
4.45 – 5.00 JONATHAN DOWNES:
Keynote Speech and Closing Remarks.

EVENING: Dinner at *The Farmer's Arms*

All Ticketholders will also get money off vouchers for local amenities. Details TBA

CFZ PEOPLE

The advent of the CFZ bloggo has not just had far reaching effects upon our publishing schedule, it has changed the human map of the CFZ. There are people who have now become part of the CFZ family, who at the beginning of the year were total strangers; Emma Biddle, Gavin Lloyd Wilson, Glen Vaudrey, Liz Clancy, Naomi and Ritchie West, and Derek Grebner, for example, and there are people whom we have known for ages like Tim Matthews and Jan Edwards who are reaching new levels of importance within the CFZ family.

Because the CFZ really is a family with all the joys, and all the sorrows that this involves.

Marjorie Braund with her unofficially adopted son At the 2007 Weird Weekend

Stuart Rickard giving piggy back to small child at the 2006 Weird Weekend

There have been a whole spate of illnesses recently, with Marjorie Braund (my adopted mum, and Dave and Ross's grandmother), our next-door neighbour Stuart Rickard (who some of you will remember from last year's Weird Weekend) and our very own Emma Biddle being taken seriously ill and having spells in hospital.

It is too early for any realistic prognoses, but one can only hope (and pray) that the outcomes will be favourable.

With the Weird Weekend approaching I always find it heartwarming how nice and unselfish some of the members of the CFZ family can be. Take

Davey and Joanne Curtis for example.

They, for the second year running, are travelling down from the north of England at their own expense, with their daughter Rosie and a car load of craft materials (also paid for at their own expense) to do monster model making and picture drawing workshops for the younger generation at this year's event.

Joanne Curtis with her young charges

It is this, I think, which makes the Weird Weekend such a unique event. After last year's conference I came in for a fair amount of criticism, some of it from people who really should have known better, for allowing children in to a "scientific symposium".

Poo!

The Weird Weekend isn't a scientific symposium. Nor is it a Fortean conference. Nor is it a village fete. Nor is it a hippy-type festival. Nor is it a traditional community event. It is none of these things, and it is all of these things and more. It is the Weird Weekend and it is completely unlike anything else that I have ever heard of.

But it works, and one of the reasons that it works is because it IS a family event. And one of the reasons why I believe that the CFZ is rapidly becoming a very important organisation over and above its position within the rarefied little world of the global cryptozoological community, is because it has something to offer for all ages. And for as long as I have anything to do with running the organisation this is the way that it will stay.

And as I don't intend to retire any time soon, please accept the fact that the child-friendly atmosphere, the surreal silliness, the comedy and the music and theatre *will* stay. Because in their own little way they are as important as the publishing, the research and the expeditions that are our bread and butter.

One of the things that I think is most important about the Weird Weekend is that it is - as far as I am aware - the only even slightly Fortean conference in the world which is not just aimed at the true believers or the Fortean faithful. The general public are welcome, and they can (and do) attend.

But we don't insult them by dumbing down the talks to make them entry-level. Our speakers are all experts in their own particular field, but they are all aware that their audience are not necessarily experts, so they tailor their talks accordingly, but they do not treat the audience – whatever age they are – as if they are stupid. And the audience - of all ages - appreciates that!

The Weird Weekend is, as I have always said, a place to see old friends and make new ones; a place where friendships and even marriages are made, where alliances forged, and working relationships formed. A place where papers are written and presented, where beer is drunk, where jokes are made and expeditions planned.

But the thing I am looking forward to most about this year's event, which is only a few months away now, is that I am going to get to meet some of my new friends who have become such important parts of the CFZ without us ever having met in the flesh.

And that is going to be a great privelige indeed.

Weird Weekend Tickets are available for £20 for the whole weekend (advance booking price) and can be bought online via paypal or by post from us here at the CFZ. Please make your cheques payable to `CFZ Trust`.

The Editor and his band of merry men welcome an exchange of correspondence on any subject of interest to readers of this magazine.

We reserve the right to edit letters and would like to stress that opinions voiced are those of the individual correspondent rather than being necessarily those of the editorial team or the Centre for Fortean Zoology. Every attempt is made not to infringe anyone's moral rights or copyright, and we apologise if we have unwittingly done so.

HOMEWRECKER

Hello,

I have sent you some information regarding yowies that I have been collecting.. (Bundle of scans of very interesting Australian newspaper reports attached)

The photos of the broken wattle tree was taken on my parents property at Kempsey N.S.W.

This tree damage was not done by livestock and it was not done by lighting either. This type of tree attack was happening on a regular basis on this property and we believe it was done by a yowie. The next door neighbour's son had seen a yowie at the back of his house and they had been seen many times on the old dairy farm just up the road from us. I have also sent you some newspaper clippings.

You might already have them but if you don't I hope you find them interesting. After seeing this large tree attacked right next to a bush track that I walked down almost every day, I would not go into the bush anymore and nobody would go into that paddock at night. Our dog had its hackles up at something in this paddock and horses would sometimes gallop out of this paddock at full speed.

There were no cattle on this property and these tree attacks were not done by horses. We have since moved from the Kempsey property. but I am currently researching yowies there and the pilliga scrub.

The following account came to us from a cousin of the people that actually had the yowie come into the house & also saw it behind their dairy shed a number of times. She also told us that her father, who grew up here around Mungay Creek saw what appeared to be a family of them in the creek when he was riding with his brothers as a boy. This lady had heard all about the yowies here first hand from her family & they were common knowledge amongst the people living here in earlier days. The instance when one entered the house occurred when the family had gone to town leaving one of the boys at home in bed as he was sick.

Apparently it came through the front door into the hall & turned & stood in his bedroom doorway looking at him. She said that he got a hell of fright & so did the yowie, which took off. After this, this particular son wouldn't stay home alone

Letters to the Editor

mentioned them or our interest in them! They were quite serious when they said it too.

That's when we asked them what they knew & she told us all about her families experience with them. If I hear of any more sightings I will let you know.

I'm not sure what year the sightings happened, but the children of the family are in their 50's now I think.

P.S The wattle tree photos were taken last December the day after it was attacked.
Regards
Michael Hardcastle.

I NEVER SAID I WAS DEEP

Jon,

Knew you were keen on mustelids, just heard from the Suffolk Wildlife Trust that polecats have re-appeared in the county of Suffolk after over a century. The last recorded polecats in Suffolk were the 38 killed in the year of 1898 by a gamekeeper in Mildenhall.

Suffolk Wildlife Trust say some typical polecat habitats have disappeared in the intervening century, but they've got a lot of roadkill to eat these days, and there's a more relaxed attitude to carnivore "management" among the conservation people these days.

They introduced Dartmoor ponies into Suffolk in the last few months - around Dunwich forest, to keep down the conifers and try to thin out the forest a bit. They put in a system of cattle grids to keep them to a certain area, but they're already being reported the wrong side of the cattle grids. I encountered one on a forest path in the middle of the night, scared the crap out of me!

Matt Salusbury

again. The yowie had left tracks behind the shed. she & her husband said, & I quote, "watch out for the yowies"! We were stunned as we had never

pictures capture the majesty of the Loch. My favourite is the last: a moody picture of the sun setting across the loch whilst in the foreground a collection of rocks and logs for a monster-like out line.

The text is hardly ground-breaking but in such a slim volume you would not expect it to be. James Carney takes a fairly sceptical view but who can blame him.

After '86 no convincing evidence has come forward. Personally I'll stick to the idea of a very big fish and no amount of plesiosaur shaped rubbers, Nessie pencil cases or plastic monsters in Tam-o-shanters will convince me otherwise. **Richard Freeman**

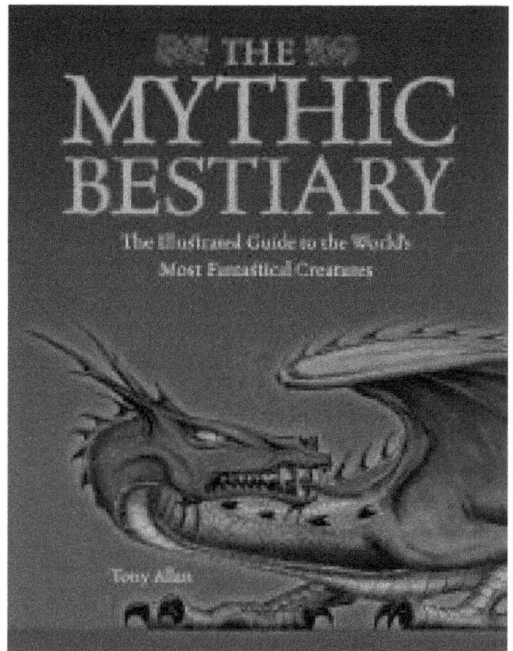

The Loch Ness Monster
Text: James Carney
Photography: Colin Baxter
Colin Baxter Photography Ltd 2009
ISBN 978-1-84107-413-9

This is a booklet in they style that was popular a few years back. The CFZ library has a number from the 70s and 80s, of varying quality. What makes this addition to the canon of Loch Ness books stand head and shoulders (or rather head and neck) above the others is the quality of the photography.

Colin Baxter is clearly a master of his trade. His

Bookshelf

The Mythic Bestiary; The Illustrated Guide to the World's Most Fantastical Creatures
Tony Allan
Duncan Baird Publishers 2008 ISBN 978-1-84483-458-7

Modern day bestiaries and monster encyclopaedias are quite common but his is a truly excellent example. Tony Allan is an historian and former series editor of Time Life books and it really shows. The book is very well written and lavishly illustrated. A number of different artist worked on this book my favourite being Tomislave Tomic who captures Japanese monsters in a 'wood cut' style similar to the celebrated 18th century Japanese artist Toriyama Sekien.

The bright green dragon on the front cover is eye catching as well. There are some less impressive pictures. The chupacabra for example, looks like a bull terrier crossed with an iguana and mothman is depicted as a man with wings rather than the inhuman form witnesses actually spoke about. But these are tiny quibbles that do not detract from a very well written and illustrated book. **Richard Freeman**

The Mystery Animals of Britain and Ireland: Kent
Neil Arnold
CFZ Press, Bideford, 2009
ISBN-13: 978-1905723362

What a smashing book! Some of us have been reading Neil Arnold's adventures in the pages of *Animals & Men* for the last decade and a half, and have been awaiting this book almost as long.

Because although, as Jon Downes claims in his introduction to the series, this book `does what it says on the tin` and gives an entertaining and in-depth view of the cryptofauna of the county sometimes known as `The Garden of England`, but it does a hell of a lot more.

This is a personal journey of discovery by one of Britain's foremost mystery animal researchers, and furthermore it is an open and honest one.

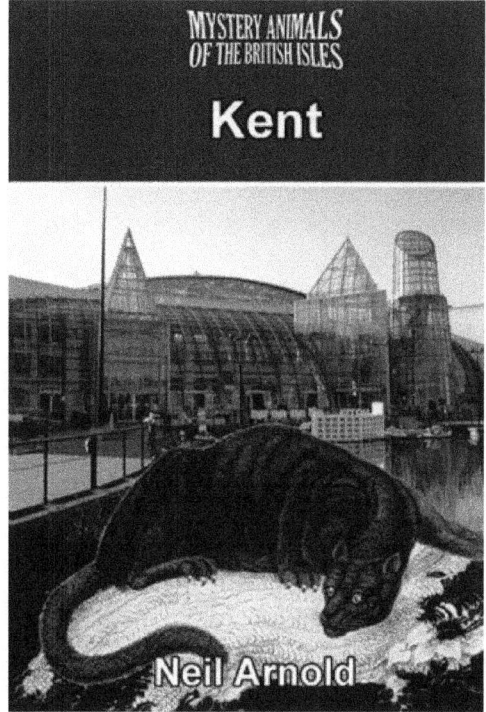

Neil has his detractors: mostly people who resent the fact that he doesn't pay lip service to their tawdry little endeavours.

He pulls no punches, but then again why should he? He is a field naturalist *par excellence* and this shows again and again in the quality of his observational writing. There should be more researchers like Neil Arnold, and less idiots always ready to take potshots at him.

However, as the CFZ have found on many occasions over the last seventeen years, the field of mystery animal research is over-populated with wannabes, and amateurs who like nothing more than to make snide attacks on those of greater ability purely because they are not able to achieve anything of substance themselves.

I recommend this book to anyone with even the slightest interest in mystery animals or the county of Kent, and look forward greatly to Neil's next book, and to the next volume in the series of Mystery Animals of Britain. **Bill Petrovic**

"How animals evolve on islands" (*Hoe Dieren op Eilanden Evolueren*),
Alexandra van der Geer, John de Vos, Michael Dermitzakis, George Lyras,
Natuurwetenschap & techniek, Diemen, Holland, 2009
ISBN 978 908571 169 8.

Island dwarves and island giants, and generally strange fossil island fauna, are the subject of this Dutch-Greek contribution to Darwin's anniversary year. Something weird happened to deer in particular when they swam by accident to Mediterranean islands – they get smaller, their antlers become truly bizarre. Some little island deer evolved blunt antlers like clubs, others developed antlers like spikes, some become so outlandish we can't even tell for certain that they were deer.

There's also a lack of biodiversity on islands is also noticeable – Cypriot three-foot pygmy elephants shared the island with mongoose-like gennet cats, a couple of species of mice, dormice and shrews, numerous pygmy hippos, bats, swans and not much else. Many churches on Cyprus displayed the miraculously preserved bones of the first Christian settlers that came to the island to flee persecution (according to Greek orthodox tradition), except that, as intrepid early 20th century British naturalist Dorothea Bate discovered, most of these relics were in fact the fossil bones of the very common pygmy hippos that populated the island millions of years before people arrived. The book has spectacular photos, the like of which you are unlikely to have seen before – a stuffed specimen on the Falkland Islands "wolf", the articulated skeleton of the fearsome Italian fossil *Deinogalerix*, a terrier-sized, bird-eating "terrible hedgehog" from the days when present-day Italy was an archipelago, newly discovered pygmy elephants of the Philippines, a never-before-seen, and rather fluffy and cute reconstruction of the five-foot pygmy mammoth *Mammuthus exilis* and some less well-known fossil and more recent island animals like the Ryuku rat.

While its authors admits that the "relationship between the island giants and the mainland giants is unclear," the book also raises the intriguing possibility that the giant tortoises of the Galapagos and the Seychelles aren't giants at all, they may turn out to be dwarf ancestors of a Volkswagen Beetle-sized mainland ancestor, something very like the huge fossil tortoise recently found on the Greek island of Lesbos. *Hoe Dieren...* also has new evidence on the impact (or lack of it) of early humans on island fauna, and pours cold water on the idea that early humans killed and cooked the last Mediterranean dwarf elephants on the island of Tilos. The black layer in the caves of Tilos, once thought to be signs of early fires, turns out to be a mineral deposit from long before people arrived.

The one serious drawback of *Hoe Dieren...* is that it's in written in Dutch, which consigns it to a tiny audience. This is especially silly, as most of the sources it cites are in English and its target audience in Holland or Belgium wouldn't have too much trouble reading a book like this – aimed at non-specialists – if it were in English. Its co-author John de Vos told me there were no plans for an English edition, but said there was a longer (unnamed) English language "scientific book" on the subject coming out next year, which we eagerly await.

Matt Salusbury

Monster Hunt; The Guide to Crypto-zoology
Rory Storm
Sterling Publishing ISBN 978-1-4027-6314-4

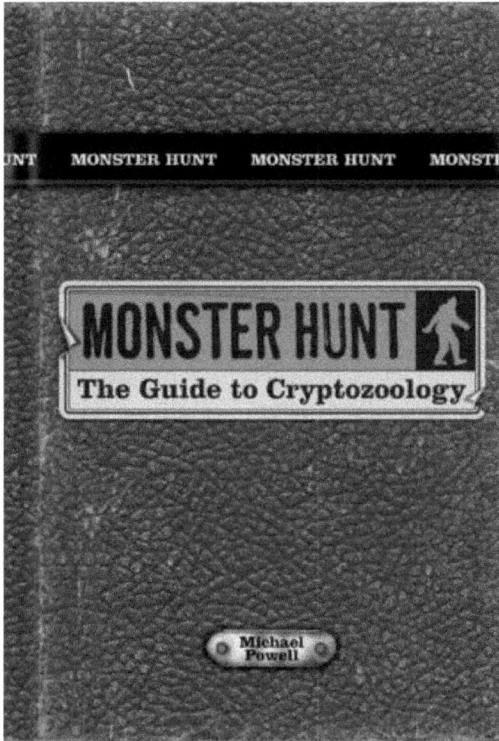

Jon Downes and I have long intended to write a children's book on cryptozoology. This book makes me want to write it even more. This is what you get when you have a book about cryptozoology that's written by a hack rather than an actual cryptozoologist. It is chock full of embarrassing mistakes. I have never believed in writing down to children, or that children's books should be any the less well researched than those for adults.

Where to start on this abomination? The one good point is that, in an original twist, the book is laid out like the journal of a Victorian explorer and the illustrations attempt to look like sepia photographs.

There is an introduction to cryptozoology, fol-lowed by a continent-by-continent look at cryptids. These are not covered in detail and there are major monitions such as the thylacine that is absent from the Australian section. Finally there is a list of cryptozoologists. The living people mentioned are infuriatingly all American. This sort of self obsessed parochialism gets up my nose. No one form the CFZ was mentioned neither was Debbie Martyr, or any of the Australian or Russian cryptozoologists. The only mention the CFZ gets in the whole book is a line in the death worm section. This was in a paragraph lifted directly from Wikipedia, word for word. This is sloppy, lazy and Wikipedia is notoriously inaccurate.

The wringing in general is extremely poor. Badly researched and full of inaccuracies. The tedious trend of explaining the Loch Ness Monster, and Mokele-mbembe as pre-historic reptiles is infuriating and flies in the face of all evidence. Ogopogo is said to have a maximum length of 20 feet. Anyone even vaguely familiar with the creature will tell you that it is far larger. Worst of all is the Santa Cruz sea monster carcass, which is well known to have been a beaked whale. In this book it is postulated to be either a plesiosaur or a basilosaur (who looked completely dissimilar any how). He accompanying drawing make it look like a mosasaur.

The reconstructions show that the artist has not read the eyewitness descriptions of the creatures. Many reconstructions look nothing like any description of the creature ever given. For example the Mongolian death worm is portrayed like a giant armoured leech, the tatzelwurm is shown with four legs when every single witness says it has two, mothman (incidentally not a cryptid) is shown as humanoid, the buru is reconstructed as a stegosaurus and Manipogo is shown as a kind of *Dunkleosteus*/crocodile hybrid.

An utter waste of time and money. If book companies want books on cryptozoology why don't they ask cryptozoologists to write the books? That way they would have some degree of accuracy. Every time I have approached a mainstream publisher I have been ignored or turned down. You can't blame me for being angry when rubbish like this sees print. **Richard Freeman**

Deep in a cave beneath
Loch Ness lives a strange
figure who steals ideas
from other magazines and
then somehow makes them
his own.

Ladies and Gentlemen, it's:

T H E

Sycophant

The paranoia of various parts of the Fortean universe never ceases to amaze me. Tim Matthews has been a member of the CFZ and fairly pivotally involved with our activities for ten years now since we first met him on the disastrous cross channel cruise with Uri Geller (to watch an eclipse).

However the fact that he was at last year's Weird Weekend (and will be at this year's) and was seen lurking menacingly at the CFZ stall during last November's Unconvention has provoked some of the great unwashed to wonder what he is up to? Is he (dread the thought) trying to destabilise cryptozoology?

Of course not.

Several weeks ago a party from the CFZ went to an auction and mini convention held by the British Cichlid Association in Redditch, a town on the outskirts of Birmingham.

On the bloggo Jon happened to mention that Tim Matthews was originally planned as part of the excursion. Its true, he was! He didn't go for reasons that are so unimportant I cannot even remember what they are, but this did put the kybosh on one of our plans. We were going to take a photograph of Timmo standing, looking sinister by a tank of tropical fish, and start a rumour that he is currently doing his best to destabilise the cichlidae.

Then, later in the summer there is a conservation day about freshwater inverts somewhere in the East Midlands, and we want to send Timmo along purely so we can claim that he is trying to destabilise water beetles.

However, whilst on the subject, the above photograph of the day Tim and Jon first met (see Jon's book *The Blackdown Mystery*) has recently resur-

faced. This is conclusive proof that both men (and presumably the CFZ as well) are financed by Mossad.

That would explain a lot.

Changing the subject entirely, Oll Lewis is having enormous problems, as we go to press, trying to purchase pheasants for the main CFZ aviary.

Describing a fruitless conversation with a well-known conservation organisation he writes:

"One example of how frustrating this conversation was when I introduced myself as 'Oliver Lewis from the Centre for Fortean Zoology' and she insisted that she didn't have anyone by that name there, and another was when she asked her colleagues if they had received any phone calls from the zoology department of Aberdeen! Well this was all starting to get stupid, I know for a fact that I don't have a strong accent, there have been times when English people have even been surprised to find out I was even Welsh, so clearly something was up with her phone line at the very least. "

Well as Oll brings up the subject of his nationality every other sentence, the old biddy must have been several rizlas short of a suspiciously long cigarette, as well as deaf.

However Oll is not alone at last! The lovely Emily Taylor (15) eponymous star of our new film *Emily and the Big Cats* which is currently in production is also Welsh, and staunchly sides with Oliver on the Celtic barriers.

Jon (pictured here with Emily's sister Jessica, was overheard muttering that if he heard another chorus of *Men of Harlech* everyone would be sacked, but there were leeks for tea and he didn't want to miss them....

THE CENTRE FOR FORTEAN ZOOLOGY

So, what is the Centre for Fortean Zoology?

We are a non profit-making organisation founded in 1992 with the aim of being a clearing house for information, and coordinating research into mystery animals around the world. We also study out of place animals, rare and aberrant animal behaviour, and Zooform Phenomena; little-understood "things" that appear to be animals, but which are in fact nothing of the sort, and not even alive (at least in the way we understand the term).

Why should I join the Centre for Fortean Zoology?

Not only are we the biggest organisation of our type in the world, but - or so we like to think - we are the best. We are certainly the only truly global Cryptozoological research organisation, and we carry out our investigations using a strictly scientific set of guidelines. We are expanding all the time and looking to recruit new members to help us in our research into mysterious animals and strange creatures across the globe. Why should you join us? Because, if you are genuinely interested in trying to solve the last great mysteries of Mother Nature, there is nobody better than us with whom to do it.

What do I get if I join the Centre for Fortean Zoology?

For £12 a year, you get a four-issue subscription to our journal *Animals & Men*. Each issue contains 60 pages packed with news, articles, letters, research papers, field reports, and even a gossip column! The magazine is A5 in format with a full colour cover. You also have access to one of the world's largest collections of resource material dealing with cryptozoology and allied disciplines, and people from the CFZ membership regularly take part in fieldwork and expeditions around the world.

How is the Centre for Fortean Zoology organized?

The CFZ is managed by a three-man board of trustees, with a non-profit making trust registered with HM Government Stamp Office. The board of trustees is supported by a Permanent Directorate of full and part-time staff, and advised by a Consultancy Board of specialists - many of whom who are world-renowned experts in their particular field. We have regional representatives across the UK, the USA, and many other parts of the world, and are affiliated with other organisations whose aims and protocols mirror our own.

I am new to the subject, and although I am interested I have little practical knowledge. I don't want to feel out of my depth. What should I do?

Don't worry. We were *all* beginners once. You'll find that the people at the CFZ are friendly and approachable. We have a thriving forum on the website which is the hub of an ever-growing electronic community. You will soon find your feet. Many members of the CFZ Permanent Directorate started off as ordinary members, and now work full-time chasing monsters around the world.

I have an idea for a project which isn't on your website. What do I do?

Write to us, e-mail us, or telephone us. The list of future projects on the website is not exhaustive. If you have a good idea for an investigation, please tell us. We may well be able to help.

How do I go on an expedition?

We are always looking for volunteers to join us. If you see a project that interests you, do not hesitate to get in touch with us. Under certain circumstances we can help provide funding for your trip. If you look on the future projects section of the website, you can see some of the projects that we have pencilled in for the next few years.

In 2003 and 2004 we sent three-man expeditions to Sumatra looking for Orang-Pendek - a semi-legendary bipedal ape. The same three went to Mongolia in 2005. All three members started off merely subscribers to the CFZ magazine.

Next time it could be you!

Project Kerinci, Sumatra - 2003
In search of the bipedal ape Orang Pendek

How is the Centre for Fortean Zoology funded?

We have no magic sources of income. All our funds come from donations, membership fees, works that we do for TV, radio or magazines, and sales of our publications and merchandise. We are always looking for corporate sponsorship, and other sources of revenue. If you have any ideas for fund-raising please let us know. However, unlike other cryptozoological organisations in the past, we do not live in an intellectual ivory tower. We are not afraid to get our hands dirty, and furthermore we are not one of those organisations where the membership have to raise money so that a privileged few can go on expensive foreign trips. Our research teams both in the UK and abroad, consist of a mixture of experienced and inexperienced personnel. We are truly a community, and work on the premise that the benefits of CFZ membership are open to all.

What do you do with the data you gather from your investigations and expeditions?

Reports of our investigations are published on our website as soon as they are available. Preliminary reports are posted within days of the project finishing.

Each year we publish a 200 page yearbook containing research papers and expedition reports too long to be printed in the journal. We freely circulate our information to anybody who asks for it.

No. Each year since 2000 we have held our annual convention - the *Weird Weekend* - in Exeter. It is three days of lectures, workshops, and excursions. But most importantly it is a chance for members of the CFZ to meet each other, and to talk with the members of the permanent directorate in a relaxed and informal setting and preferably with a pint of beer in one hand. Since 2006 - the *Weird Weekend* has been bigger and better and held in the idyllic rural location of Woolsery in North Devon. The 2008 event will be held over the weekend 15-17 August.

Since relocating to North Devon in 2005 we have become ever more closely involved with other community organisations, and we hope that this trend will continue. We also work closely with Police Forces across the UK as consultants for animal mutilation cases, and we intend to forge closer links with the coastguard and other community services. We want to work closely with those who regularly travel into the Bristol Channel, so that if the recent trend of exotic animal visitors to our coastal waters continues, we can be out there as soon as possible.

We are building a Visitor's Centre in rural North Devon. This will not be open to the general public, but will provide a museum, a library and an educational resource for our members (currently over 400) across the globe. We are also planning a youth organisation which will involve children and young people in our activities. We work closely with *Tropiquaria* - a small zoo in north Somerset, and have several exciting conservation projects planned.

Apart from having been the only Fortean Zoological organisation in the world to have consistently published material on all aspects of the subject for over a decade, we have achieved the following concrete results:

- Disproved the myth relating to the headless so-called sea-serpent carcass of Durgan beach in Cornwall 1975
- Disproved the story of the 1988 puma skull of Lustleigh Cleave
- Carried out the only in-depth research ever into the mythos of the Cornish Owlman
- Made the first records of a tropical species of lamprey
- Made the first records of a luminous cave gnat larva in Thailand.
- Discovered a possible new species of British mammal - the beech marten.
- In 1994-6 carried out the first archival fortean zoological survey of Hong Kong.
- In the year 2000, CFZ theories where confirmed when an entirely new species of lizard was found resident in Britain.
- Identified the monster of Martin Mere in Lancashire as a giant wels catfish
- Expanded the known range of Armitage's skink in the Gambia by 80%
- Obtained photographic evidence of the remains of Europe's largest known pike
- Carried out the first ever in-depth study of the *ninki-nanka*
- Carried out the first attempt to breed Puerto Rican cave snails in captivity
- Were the first European explorers to visit the `lost valley` in Sumatra
- Published the first ever evidence for a new tribe of pygmies in Guyana
- Published the first evidence for a new species of caiman in Guyana

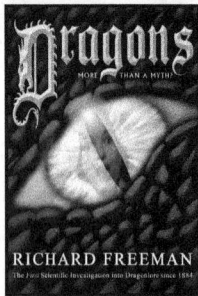

Other books available from
CFZ PRESS

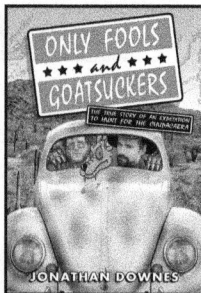

Other books available from
CFZ PRESS

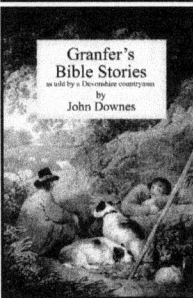

Other books available from
CFZ PRESS

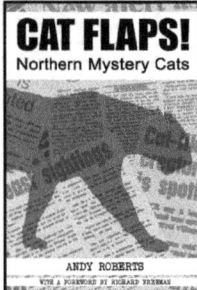

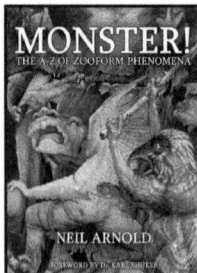

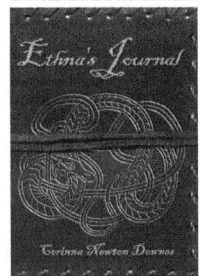

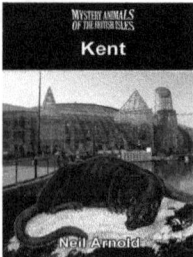